Internet of Things - New Trends, Challenges and Hurdles

Edited by Manuel Domínguez-Morales,
Ángel Varela-Vaca
and Lourdes Miró-Amarante

Published in London, United Kingdom

Internet of Things - New Trends, Challenges and Hurdles
http://dx.doi.org/10.5772/intechopen.97988
Edited by Manuel Domínguez-Morales, Ángel Varela-Vaca and Lourdes Miró-Amarante

Contributors
Seifeddine Messaoud, Rim Amdouni, Mohamed Ali Hajjaji, Abdelatif Mtibaa, Mohamed Atri, Adnen Albouchi, Francisco Luna-Perejón, Javier Civit-Masot, Luis Muñoz-Saavedra, Elena Escobar-Linero, Mnar Alnaghes, Nickolas Falkner, Hong Shen, Linos Nchena, Martin Tomášek, Manuel Jesus Dominguez-Morales, Ángel Varela-Vaca, Lourdes Miró-Amarante, Miguel Civit, Francisco Cuadrado, Charles Tijus, María José Escalona

First published in London, United Kingdom, 2023 by IntechOpen
IntechOpen is the global imprint of INTECHOPEN LIMITED, registered in England and Wales, registration number: 11086078, 5 Princes Gate Court, London, SW7 2QJ, United Kingdom

British Library Cataloguing-in-Publication Data
A catalogue record for this book is available from the British Library

Additional hard and PDF copies can be obtained from orders@intechopen.com

Internet of Things - New Trends, Challenges and Hurdles
Edited by Manuel Domínguez-Morales, Ángel Varela-Vaca and Lourdes Miró-Amarante
p. cm.
Print ISBN 978-1-80356-035-9
Online ISBN 978-1-80356-036-6
eBook (PDF) ISBN 978-1-80356-037-3

Meet the editors

Manuel Domínguez-Morales obtained a bachelor's degree in Computer Science Engineering in 2008, a master's degree in Software Engineering and Technology in 2009, a master's degree in Computer and Networks Engineering in 2014, and a Ph.D. in Industrial Informatics in 2014, all from the University of Seville, Spain. He is an associate professor in the Architecture and Computer Technology Department, at the University of Seville, and a member of the university's Robotics and Computer Technology Lab. He has authored and co-authored more than 100 international publications including conferences and journal articles, with 30 manuscripts published in JCR-indexed journals.

Ángel Varela-Vaca obtained a BS in Computer Engineering in 2008, an MSc in Software Engineering and Technology in 2009, and a Ph.D. in Computer Science (with honors) in 2013, all from the University of Seville, Spain. He is an associate professor in the Languages and System Informatics Department, at the University of Seville, and belongs to the IDEA Research Group. Dr. Varela-Vaca has led various private projects, participated in several public research projects, and published several high-impact papers. He was nominated as a member of program committees such as the 25th International Conference on Information Systems Development (ISD 2016), 2017 Business Process Management (BPM) workshops, 24th ACM International Systems and Software Product Line Conference (SPLC 2020), 8th International Symposium on Data-Driven Process Discovery and Analysis (SIMPDA 2018), and others. He has been a reviewer for many international journals, including the *Journal of Supercomputing* and *IEEE Access*, among others.

Lourdes Miró-Amarante obtained a bachelor's degree in Computer Science Engineering in 1999 and a Ph.D. in Industrial Informatics in 2013, both from the University of Seville, Spain. She is an assistant professor in the Computer Architecture and Technology Department, at the University of Seville, and a member of the university's Robotics and Computer Technology Lab. She has more than 3500 teaching hours in technical degrees and 200 in master's degrees. She coordinated the Guidance and Tutorial Action Plan (ETSII) in 2014–2016, and the master's degree in Biomedical Engineering and Digital Health in 2019–2022. She has been part of the Advisory Group for Pedagogical Support of Educational Technologies and Digital Resources since May 2020. From 2017 to 2021, Dr. Miro-Amarante was director of the secretariat for Teaching Innovation and Digitization, at the International University of Andalusia. Since 2022, she has been deputy director for Students, Innovation and Social Responsibility (ETSII).

Contents

Preface

The Internet of Things (IoT) refers to a network of physical devices that communicate over the Internet using sensors, software, and other technologies. It has numerous applications in fields ranging from industry (e.g., Industry 4.0) to health care (e.g., telemonitoring, mobile health).

When talking about IoT, it is important to focus on two fundamental aspects: the communication mechanism and the connected devices. The IoT is a multidisciplinary field that integrates both hardware and software.

The expansion in the use of IoT technology in society has not been as widespread as it has been in the field of research. Therefore, although the technology is currently at a high level of maturity, there is still a lot of room for expansion and application in society. For this reason, it is essential that the research studies carried out serve as a transfer to society and allow for the implementation of the IoT in a real environment. Even though the IoT is already being used in everyday life in the form of smart applications, research in this area focuses on more risky solutions.

This book presents several recent studies on the IoT that have great potential to be extrapolated to the real world in the future. We hope that the information contained herein will inspire future advances in the area.

Manuel Domínguez-Morales, Ángel Varela-Vaca and Lourdes Miró-Amarante
E.T.S. Ingeniería Informática,
University of Seville,
Seville, Spain

Section 1

Introduction and Future Perspectives

Chapter 1

Introductory Chapter: An Overview to the Internet of Things

Manuel Domínguez-Morales, Ángel Varela-Vaca and Lourdes Miró-Amarante

1. Introduction

The Internet of Things (IoT) refers to the process of connecting everyday physical objects to the internet, from common household objects (lighting, appliances, etc.) to healthcare assets (such as medical devices), as well as wearables, smart devices and even smart cities.

These IoT-connected physical objects are visible within the created network itself, allowing them to be consulted and/or acted upon.

The great advantage of the IoT, which leads to its enormous importance today, centres on the ease of connecting new objects to this network. The interest in this technology is being increased year by year, as shown in **Figure 1**.

A few years ago, in order to connect a device to the network, it was necessary to deploy a multi-layered infrastructure to access its information. Nowadays, however, there are open-access projects that present a free and extensive network where end users can directly connect their objects (only needing a connection modem in the object itself).

With multiple connected objects over a large area, there is great potential for projects that focus on the population's well-being when applied to smart cities. This opens up endless possibilities, but not without challenges and concerns. Many of the latter focus on the devices and network's security and devices and how a malicious user can alter the information or undermine privacy.

All these issues and possibilities are addressed in the various chapters of this book, which attempt to cover all areas of the IoT.

So, the main aim of this introductory chapter is to serve as a justification for the book itself, presenting hard facts and data that prove the evolution of the use and deployment of IoT systems in society. To this end, a literature review will be carried out to show the increase in publications related to the subject in recent years.

2. Trend analysis

The methodology used corresponds to the classical systematic review process. The keywords used for the search process are "Internet of Things" and "IoT", including the operand "OR" between both. In order to observe the trend, the last 20 complete years are taken into account (from 2001 to 2021). Finally, the search engines used for it are Google Scholar, IEEE Xplore and Scopus. With the information obtained, the criteria used to analyse the works is mainly the applied field.

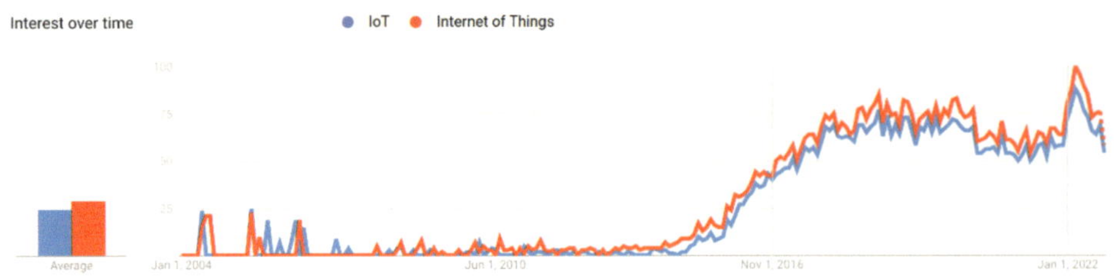

Figure 1.
Interest over time in the terms "IoT" and "Internet of Things" from 1st January 2004 to date (obtained from Google Trends). A strong increase can be observed in 2016 and in the beginning of 2022.

All the works found are used to obtain the distribution per year and observe the tendency. However, not all these works are analysed deeply to find their topic because of the great number of works found. Instead, we analyse only a subset of the most-cited works from each year.

The search results show a total of 339.804 works published between 1 January 2001 and 31 December 2021. The evolution of these publications for each year can be seen in **Figure 2**. It can be seen that the number of publications between 2001 and 2008 is not more than 100. The increase was maintained in subsequent years, but it was not until 2016 that a breakpoint was observed, with the number of papers doubling that of the previous year. This point coincides with the annotation observed in **Figure 1**.

From 2016, there was an exponential increase until 2020, when stagnation is observed (presumably due to the pandemic) with a subsequent upturn in 2021.

As a result, it can be theorised that the trend in interest and use of IoT technologies has passed its exponential growth stage and is in the maintenance stage. It is at this point where it can be theorised that the research linked to this field is in its maturity stage, and therefore, we are in an ideal position to be able to publish a book of these characteristics.

In order to analyse the topic distribution, the most cited works from each year are extracted using the next criteria:

- From 2001 to 2008: in this period, there were less than 100 works per year (92 in 2008), so we extract the 10% most cited works for each year. In total, we obtain 25 works in this period.

- From 2009 to 2011: in this period, the number of works per year varied between 100 and 1000. As there is a big variation between these years, we extract the

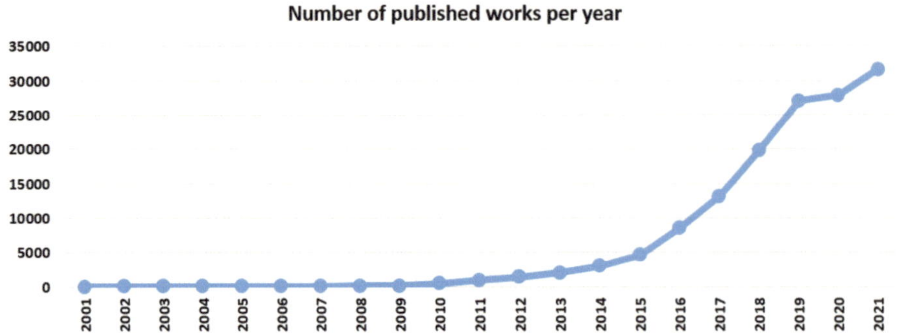

Figure 2.
Number of works published each year from 2001 to 2021 using the search phrase "IoT" OR "Internet of Things".

3% most cited works for each year with a minimum of 10. In total, we obtain 50 works in this period.

- From 2012 to 2015: in this period, the number of works per year varied between 1000 and 5000. For this case, we extract the 1% most cited works per year with a maximum of 40. In total, we obtain 102 works in this period.

- From 2016 to 2021: this is the period with the biggest number of published works (from 13 to 31 K works), so we need to reduce the number of analysed works in order to simplify the evaluation stage. So, we extract the 0.5% most cited works per year during this period. In total, we obtain 636 works in this period.

Finally, we obtained 813 works to be analysed. This amount of work is considerable and needs to be reduced. By discarding those works not published in international journals, the number of works is reduced to 391. Finally, discarding those published in non-JCR journals, the total amount of works is almost halved, obtaining a final number of 192 works.

With this final amount of work, the main topic distribution will be analysed. We will start by including the selected papers for the entire period (from 2001 to 2021).

If we look at the distribution of papers by each of the areas of interest (see **Figure 3**), we can see a high percentage of papers related with the field of computing (including those related with communications and security), which seems logical given the nature of the technology. In the second position is the field of Engineering, with 27% of the references observed. This is followed by pure sciences and health sciences with 16 and 15%, respectively. Lastly, the area least related with the subject of this book (social sciences) obtained 6%.

Secondly, only the selected set of works within the period from 2016 to 2021 (the period of exponential increase and stabilisation) will be analysed (see **Figure 4**). Analysing the results obtained, a very similar distribution to that obtained for the whole period can be observed. The only difference is that the first two branches (computer

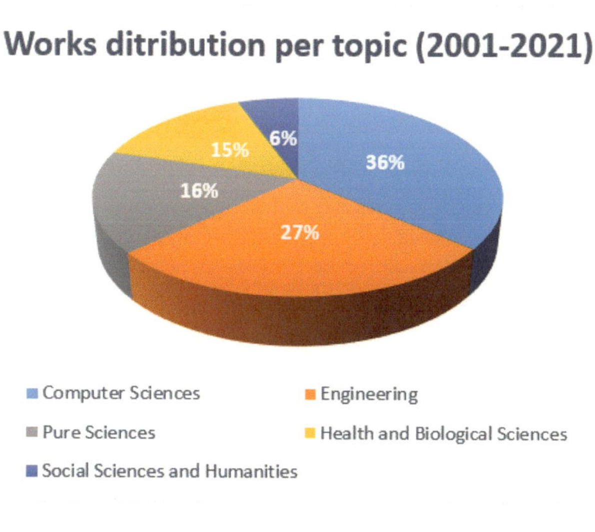

Figure 3.
Number of works published between 2001 and 2021 divided thematically.

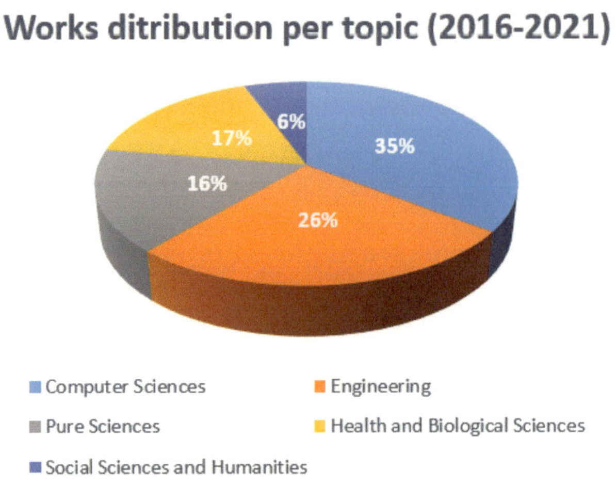

Figure 4.
Number of works published between 2016 and 2021 divided thematically.

Year	#	Work/s
2004	1	[1]
2005	1	[2]
2006	2	[3, 4]
2007	2	[5, 6]
2008	3	[7–9]
2009	4	[10–13]
2010	4	[14–17]
2011	9	[18–26]
2012	6	[27–33]
2013	6	[34–39]
2014	8	[40–47]
2015	10	[48–57]
2016	9	[58–66]
2017	12	[67–78]
2018	20	[79–98]
2019	24	[99–122]
2020	30	[107, 123–150]
2021	40	[151–188]

Table 1.
Selected works evaluated year by year.

science and engineering) slightly reduce their number in favour of health sciences (which increases from 15–17%).

In summary, therefore, it can be seen that we are currently in a period of technological maturity after a few years of exponential growth in the number of jobs. And,

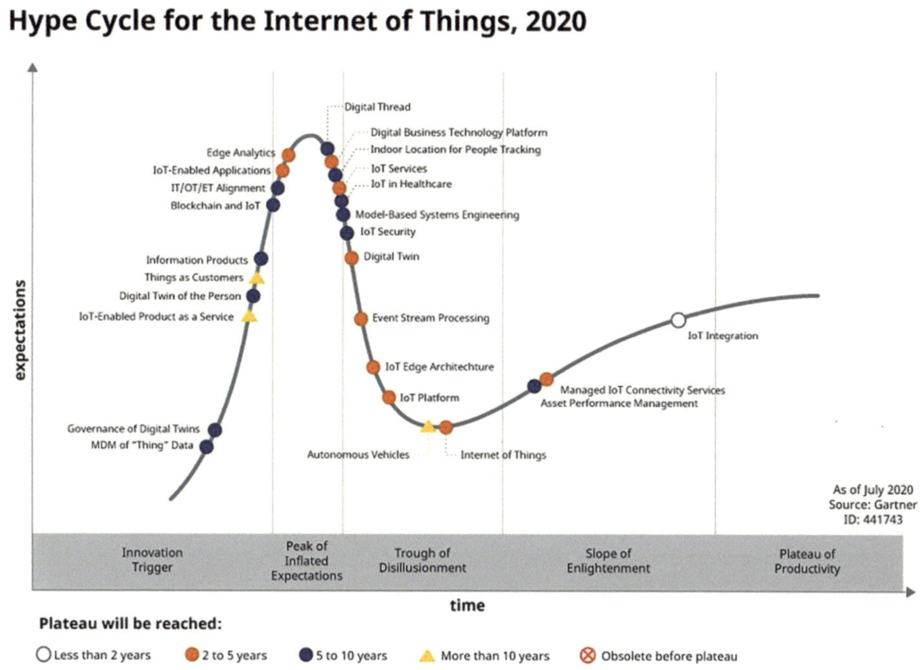

Figure 5.
Gartner's Hype Cycle of "Internet of Things".

with respect to the areas, a similar distribution is maintained throughout the period, although a continuous growth is observed in the field of health sciences.

These results have highlighted the importance and evolution of the Internet of Things in recent years. A significant increase in the number of publications has been observed since 2016, coinciding with the search trends provided by Google Trends.

This upward trend continues to increase exponentially until it stagnates in 2020, something that can also be seen in the search trends.

The summary of the most-representative works evaluated is presented in **Table 1**.

These data are directly related to the latest Gartner Hype Cycle of Internet of Things (published in 2020). It can be seen in **Figure 5** how the initial themes linked to the Internet of Things (including IoT Edge and IoT Platform) have already passed the crest of the wave and are in decline: these technologies were the fruit of the first upturn in 2016 (when both were at the crest of the wave). However, it can be seen that the technologies currently at their peak include those related with IoT in healthcare and smart homes, which may justify the increase in the proportion of publications in the health sciences in recent years.

Therefore, IoT systems and technologies have passed their initial curve of novelty and technology evolution and are now mature enough to be able to find innovative and useful work already implemented in society. It is therefore an ideal time to produce this book.

Author details

Manuel Domínguez-Morales[1*], Ángel Varela-Vaca[2] and Lourdes Miró-Amarante[1]

1 Architecture and Computer Technology Department, University of Seville, Seville, Spain

2 Languages and Computer Systems Department, University of Seville, Seville, Spain

*Address all correspondence to: mjdominguez@us.es

References

[1] Gershenfeld N. The Internet of Things. Scientific American. 2004;**291**(4):76-81

[2] Srivastava L. The Internet of Things. International Telecommunication Union. Tech. Rep, 7. 2005. p. 212.

[3] Ning HS. Research on China Internet of Things' services and management. Acta Electonica Sinica. 2006;**34**(S1):2514

[4] Tuters M. Beyond locative media: Giving shape to the Internet of Things. Leonardo. 2006;**39**(4):357-363

[5] Gao J. RFID coding, name and information service for Internet of Things. In: IET Conference on Wireless, Mobile and Sensor Networks. 2007. pp. 36-39

[6] Ziekow H. In-network event processing in a peer to peer broker network for the Internet of Things. In OTM Confederated International Conferences on the Move to Meaningful Internet Systems. 2007. pp. 970-979

[7] Hompel M. Cellular transport systems-making things move in the "Internet of Things". IT-MUNCHEN. 2008;**50**(1):59

[8] Quack T. Object Recognition for the Internet of Things. In: The Internet of Things. Berlin, Heidelberg: Springer; 2008. pp. 230-246

[9] Wu Y. Realizing the Internet of Things in Service-Centric Environments. In ICSOC PhD Symposium. 2008

[10] Broll G. Perci: Pervasive service interaction with the Internet of Things. IEEE Internet Computing. 2009;**13**(6):74-81

[11] Kortuem G. Smart objects as building blocks for the Internet of Things. IEEE Internet Computing. 2009;**14**(1):44-51

[12] Welbourne E. Building the Internet of Things using RFID: The RFID ecosystem experience. IEEE Internet Computing. 2009;**13**(3):48-55

[13] Wright M. Deeply embedded devices: The Internet of Things-design solution-microchip technology-implementing smart IP connectivity schemes will improve efficiency and deliver better performance, which becomes more critical as the Internet of connected things approaches a billion by 2011. Electronic Design. 2009;**57**(19):41

[14] Hong S. SNAIL: An IP-based wireless sensor network approach to the Internet of Things. IEEE Wireless Communications. 2010;**17**(6):34-42

[15] Weber RH. Internet of Things–New security and privacy challenges. Computer Law & Security Review. 2010;**26**(1):23-30

[16] Zuehlke D. Smart Factory—Towards a factory-of-things. Annual Reviews in Control. 2010;**34**(1):129-138

[17] Atzori L, Iera A, Morabito G. The internet of things: A survey. Computer Networks. 2010;**54**(15):2787-2805

[18] Alam S. Interoperability of security-enabled Internet of Things. Wireless Personal Communications. 2011;**61**(3):567-586

[19] Atzori L. Siot: Giving a social structure to the Internet of Things. IEEE Communications Letters. 2011;**15**(11):1193-1195

[20] Foschini L. M2M-based metropolitan platform for IMS-enabled road traffic management in IoT. IEEE Communications Magazine. 2011;**49**(11):50-57

[21] Kiritsis D. Closed-loop PLM for intelligent products in the era of the Internet of Things. Computer-Aided Design. 2011;**43**(5):479-501

[22] Li X. Smart community: An Internet of Things application. IEEE Communications Magazine. 2011;**49**(11):68-75

[23] Ning H. Future Internet of Things architecture: Like mankind neural system or social organization framework? IEEE Communications Letters. 2011;**15**(4):461-463

[24] Roman R. Securing the Internet of Things. Computer. 2011;**44**(9):51-58

[25] Xu LD. Information architecture for supply chain quality management. International Journal of Production Research. 2011;**49**(1):183-198

[26] Zhou L. Multimedia traffic security architecture for the Internet of Things. IEEE Network. 2011;**25**(3):35-40

[27] Barnaghi P. Semantics for the Internet of Things: Early progress and back to the future. International journal on Semantic Web and Information Systems (IJSWIS). 2012;**8**(1):1-21

[28] Bormann C. Coap: An application protocol for billions of tiny internet nodes. IEEE Internet Computing. 2012;**16**(2):62-67

[29] Chen M. Machine-to-machine communications: Architectures, standards and applications. KSII Transactions on Internet and Information Systems (TIIS). 2012;**6**(2):480-497

[30] Gomez C. Overview and evaluation of bluetooth low energy: An emerging low-power wireless technology. Sensors. 2012;**12**(9):11734-11753

[31] Hancke GP. The role of advanced sensing in smart cities. Sensors. 2012;**13**(1):393-425

[32] Li S. Compressed sensing signal and data acquisition in wireless sensor networks and Internet of Things. IEEE Transactions on Industrial Informatics. 2012;**9**(4):2177-2186

[33] Watteyne T. OpenWSN: A standards-based low-power wireless development environment. Transactions on Emerging Telecommunications Technologies. 2012;**23**(5):480

[34] Kelly SDT. Towards the implementation of IoT for environmental condition monitoring in homes. IEEE Sensors Journal. 2013;**13**(10):3846-3853

[35] Lazarescu MT. Design of a WSN platform for long-term environmental monitoring for IoT applications. IEEE Journal on Emerging and Selected Topics in Circuits and Systems. 2013;**3**(1):45-54

[36] Lee J. Recent advances and trends in predictive manufacturing systems in big data environment. Manufacturing Letters. 2013;**1**(1):38-41

[37] Liu V. Ambient backscatter: Wireless communication out of thin air. ACM SIGCOMM Computer Communication Review. 2013;**43**(4):39-50

[38] Raza S. SVELTE: Real-time intrusion detection in the Internet of Things. Ad Hoc Networks. 2013;**11**(8):2661-2674

[39] Wallgren L. Routing attacks and countermeasures in the RPL-based Internet of Things. International Journal of Distributed Sensor Networks. 2013;**9**(8):794326

[40] Bonomi F. Fog computing: A platform for Internet of Things and analytics. In: Big Data and Internet of Things. Springer; 2014. pp. 169-186. Available from: https://link.springer.com/chapter/10.1007/978-3-319-05029-34_7

[41] Fan YJ. IoT-based smart rehabilitation system. IEEE Transactions on Industrial Informatics. 2014;**10**(2):1568-1577

[42] Jin J. An information framework for creating a smart city through Internet of Things. IEEE Internet of Things Journal. 2014;**1**(2):112-121

[43] Perera C. Sensing as a service model for smart cities supported by Internet of Things. Transactions on Emerging Telecommunications Technologies. 2014;**25**(1):81-93

[44] Schreier G. The Internet of Things for personalized health. Studies in Health Technology and Informatics. 2014;**200**:22-31

[45] Tao F. CCIoT-CMfg: Cloud computing and Internet of Things-based cloud manufacturing service system. IEEE Transactions on Industrial Informatics. 2014;**10**(2):1435-1442

[46] Wunder G. 5GNOW: Non-orthogonal, asynchronous waveforms for future mobile applications. IEEE Communications Magazine. 2014;**52**(2):97-105

[47] Jazdi N. Cyber physical systems in the context of Industry 4.0. In: 2014 IEEE International Conference on Automation, Quality and Testing, Robotics. IEEE; 2014. pp. 1-4

[48] Catarinucci L. An IoT-aware architecture for smart healthcare systems. IEEE Internet of Things Journal. 2015;**2**(6):515-526

[49] Kamalinejad P. Wireless energy harvesting for the Internet of Things. IEEE Communications Magazine. 2015;**53**(6):102-108

[50] Gretzel U. Smart tourism: Foundations and developments. Electronic Markets. 2015;**25**(3):179-188

[51] Haas H. What is lifi? Journal of Lightwave Technology. 2015;**34**(6):1533-1544

[52] Islam SR. The Internet of Things for health care: A comprehensive survey. IEEE Access. 2015;**3**:678-708

[53] Ma X. Large-scale transportation network congestion evolution prediction using deep learning theory. PLoS One. 2015;**10**(3):e0119044

[54] Niu S. A universal self-charging system driven by random biomechanical energy for sustainable operation of mobile electronics. Nature Communications. 2015;**6**(1):1-8

[55] Satyanarayanan M. Edge analytics in the Internet of Things. IEEE Pervasive Computing. 2015;**14**(2):24-31

[56] Wang S. Triboelectric nanogenerators as self-powered active sensors. Nano Energy. 2015;**11**:436-462

[57] Zhu C. Green Internet of Things for smart world. IEEE Access. 2015;**3**:2151-2162

[58] Augustin A. A study of LoRa: Long range & low power networks for the Internet of Things. Sensors. 2016;**16**(9):1466

[59] Centenaro M. Long-range communications in unlicensed bands: The rising stars in the IoT and smart city scenarios. IEEE Wireless Communications. 2016;**23**(5):60-67

[60] Christidis K. Blockchains and smart contracts for the Internet of Things. IEEE Access. 2016;**4**:2292-2303

[61] Dastjerdi AV. Fog computing: Helping the Internet of Things realize its potential. Computer. 2016;**49**(8): 112-116

[62] Dimitrov DV. Medical Internet of Things and big data in healthcare. Healthcare Informatics Research. 2016;**22**(3):156-163

[63] Palattella MR. Internet of Things in the 5G era: Enablers, architecture, and business models. IEEE Journal on Selected Areas in Communications. 2016;**34**(3):510-527

[64] Rathore MM. Urban planning and building smart cities based on the Internet of Things using big data analytics. Computer Networks. 2016;**101**:63-80

[65] Roblek V. A complex view of industry 4.0. SAGE Open. 2016;**6**(2):2158244016653987

[66] Yang Z. An IoT-cloud based wearable ECG monitoring system for smart healthcare. Journal of Medical Systems. 2016;**40**(12):1-11

[67] Gravina R. Multi-sensor fusion in body sensor networks: State-of-the-art and research challenges. Information Fusion. 2017;**35**:68-80

[68] Gupta H. iFogSim: A toolkit for modeling and simulation of resource management techniques in the Internet of Things, edge and fog computing environments. Software: Practice and Experience. 2017;**47**(9):1275-1296

[69] Kolias C. DDoS in the IoT: Mirai and other botnets. Computer. 2017;**50**(7):80

[70] Lin J. A survey on Internet of Things: Architecture, enabling technologies, security and privacy, and applications. IEEE Internet of Things Journal. 2017;**4**(5):1125-1142

[71] Motlagh NH. UAV-based IoT platform: A crowd surveillance use case. IEEE Communications Magazine. 2017;**55**(2):128-134

[72] Mozaffari M. Mobile unmanned aerial vehicles (UAVs) for energy-efficient Internet of Things communications. IEEE Transactions on Wireless Communications. 2017;**16**(11):7574-7589

[73] Satyanarayanan M. The emergence of edge computing. Computer. 2017;**50**(1):30-39

[74] Tao F. Digital twin shop-floor: A new shop-floor paradigm towards smart manufacturing. IEEE Access. 2017;**5**:20418-20427

[75] Wang YPE. A primer on 3GPP narrowband Internet of Things. IEEE Communications Magazine. 2017;**55**(3):117-123

[76] Kshetri N. Can blockchain strengthen the Internet of Things? IT Professional. 2017;**19**(4):68-72

[77] Schulz P. Latency critical IoT applications in 5G: Perspective on the design of radio interface and network architecture. IEEE Communications Magazine. 2017;**55**(2):70-78

[78] Huh S, Cho S, Kim S. Managing IoT devices using blockchain platform. In: 2017 19th International Conference on Advanced Communication Technology (ICACT)464-467. IEEE; 2017

[79] Aazam M. Deploying fog computing in industrial Internet of Things and industry 4.0. IEEE Transactions on Industrial Informatics. 2018;**14**(10):4674-4682

[80] Boyes H. The industrial Internet of Things (IIoT): An analysis framework. Computers in Industry. 2018;**101**:1-12

[81] Diro AA. Distributed attack detection scheme using deep learning approach for Internet of Things. Future Generation Computer Systems. 2018;**82**:761-768

[82] Doshi R. Machine learning ddos detection for consumer Internet of Things devices. In 2018 IEEE Security and Privacy Workshops (SPW). May 2018. pp. 29-35

[83] Elhoseny M. Secure medical data transmission model for IoT-based healthcare systems. IEEE Access. 2018;**6**:20596-20608

[84] Goap A. An IoT based smart irrigation management system using machine learning and open source technologies. Computers and Electronics in Agriculture. 2018;**155**:41

[85] Hammi MT. Bubbles of trust: A decentralized blockchain-based authentication system for IoT. Computers & Security. 2018;**78**:126-142

[86] Khan MA. IoT security: Review, blockchain solutions, and open challenges. Future Generation Computer Systems. 2018;**82**:395-411

[87] Li C. Analogue signal and image processing with large memristor crossbars. Nature Electronics. 2018;**1**(1):52-59

[88] Masood A. Computer-assisted decision support system in pulmonary cancer detection and stage classification on CT images. Journal of Biomedical Informatics. 2018;**79**:117-128

[89] Meidan Y. N-baiot—Network-based detection of IoT botnet attacks using deep autoencoders. IEEE Pervasive Computing. 2018;**17**(3):12-22

[90] Mohamed P. Maintaining security and privacy in health care system using learning based deep-Q-networks. Journal of Medical Systems. 2018;**42**(10):1-10

[91] Novo O. Blockchain meets IoT: An architecture for scalable access management in IoT. IEEE Internet of Things Journal. 2018;**5**(2):1184-1195

[92] Pan J. Future edge cloud and edge computing for Internet of Things applications. IEEE Internet of Things Journal. 2018;**5**(1):439-449

[93] Rahmani AM. Exploiting smart e-Health gateways at the edge of healthcare Internet-of-Things: A fog computing approach. Future Generation Computer Systems. 2018;**78**:641-658

[94] Sadowski S. Rssi-based indoor localization with the Internet of Things. IEEE Access. 2018;**6**:30149-30161

[95] Sharma PK. A software defined fog node based distributed blockchain cloud architecture for IoT. IEEE Access. 2018;**6**:115-124

[96] Stergiou C. Secure integration of IoT and cloud computing. Future Generation Computer Systems. 2018;**78**:964-975

[97] Tao F. Data-driven smart manufacturing. Journal of Manufacturing Systems. 2018;**48**:157-169

[98] Verma P. Fog assisted-IoT enabled patient health monitoring in smart homes. IEEE Internet of Things Journal. 2018;**5**(3):1789-1796

[99] Behera TM. Residual energy-based cluster-head selection in WSNs for IoT application. IEEE Internet of Things Journal. 2019;**6**(3):5132-5139

[100] Chaabouni N. Network intrusion detection for IoT security based on learning techniques. IEEE Communications Surveys & Tutorials. 2019;**21**(3):2671-2701

[101] Cheng N. Space/aerial-assisted computing offloading for IoT applications: A learning-based approach. IEEE Journal on Selected Areas in Communications. 2019;**37**(5):1117-1129

[102] Cui Z. Optimal LEACH protocol with modified bat algorithm for big data sensing systems in Internet of Things. Journal of Parallel and Distributed Computing. 2019;**132**:217

[103] Dwivedi AD. A decentralized privacy-preserving healthcare blockchain for IoT. Sensors. 2019;**19**(2):326

[104] Hasan M. Attack and anomaly detection in IoT sensors in IoT sites using machine learning approaches. Internet of Things. 2019;**7**:100059

[105] Huang L. Deep reinforcement learning for online computation offloading in wireless powered mobile-edge computing networks. IEEE Transactions on Mobile Computing. 2019;**19**(11):2581-2593

[106] Khanna A. Evolution of Internet of Things (IoT) and its significant impact in the field of precision agriculture. Computers and Electronics in Agriculture. 2019;**157**:218-231

[107] Liu X. NOMA-based resource allocation for cluster-based cognitive industrial Internet of Things. IEEE Transactions on Industrial Informatics. 2019;**16**(8):5379-5388

[108] Mekki K. A comparative study of LPWAN technologies for large-scale IoT deployment. ICT Express. 2019;**5**(1):1-7

[109] Min M. Learning-based computation offloading for IoT devices with energy harvesting. IEEE Transactions on Vehicular Technology. 2019;**68**(2):1930-1941

[110] Moustafa N. An ensemble intrusion detection technique based on proposed statistical flow features for protecting network traffic of Internet of Things. IEEE Internet of Things Journal. 2019;**6**(3):4815-4830

[111] Muangprathub J. IoT and agriculture data analysis for smart farm. Computers and Electronics in Agriculture. 2019;**156**:467-474

[112] Mutlag AA. Enabling technologies for fog computing in healthcare IoT systems. Future Generation Computer Systems. 2019;**90**:62-78

[113] Neshenko N. Demystifying IoT security: An exhaustive survey on IoT vulnerabilities and a first empirical look on Internet-scale IoT exploitations. IEEE Communications Surveys & Tutorials. 2019;**21**(3):2702-2733

[114] Ning Z. A cooperative partial computation offloading scheme for mobile edge computing enabled Internet of Things. IEEE Internet of Things Journal. 2019;**6**(3):4804-4814

[115] Oh JY. Second skin enabled by advanced electronics. Advanced Science. 2019;**6**(11):1900186

[116] Shen M. Privacy-preserving support vector machine training over blockchain-based encrypted IoT data in smart cities. IEEE Internet of Things Journal. 2019;**6**(5):7702

[117] Sivanathan A. Classifying IoT devices in smart environments using network traffic characteristics. IEEE Trans. on Mobile Computing. 2019;**18**(8):1745-1759

[118] Ting DS. Deep learning in ophthalmology: The technical and clinical considerations. Progress in Retinal and Eye Research. 2019;**72**:100759

[119] Wang ZL. Entropy theory of distributed energy for Internet of Things. Nano Energy. 2019;**58**:669-672

[120] Yang Y. Privacy-preserving smart IoT-based healthcare big data storage

and self-adaptive access control system. Information Sciences. 2019;**479**: 567-592

[121] Zamora-Izquierdo MA. Smart farming IoT platform based on edge and cloud computing. Biosystems Engineering. 2019;**177**:4-17

[122] Zhang Y. Smart contract-based access control for the Internet of Things. IEEE Internet of Things Journal. 2018;**6**(2):1594-1605

[123] Aceto G. Industry 4.0 and health: Internet of Things, big data, and cloud computing for healthcare 4.0. Journal of Industrial Information Integration. 2020;**18**:100129

[124] Chen G. Smart textiles for electricity generation. Chemical Reviews. 2020;**120**(8):3668-3720

[125] Cui Z. Personalized recommendation system based on collaborative filtering for IoT scenarios. IEEE Transactions on Services Computing. 2020;**13**(4):685-695

[126] Hossein Motlagh N. Internet of Things (IoT) and the energy sector. Energies. 2020;**13**(2):494

[127] Lu Y. Blockchain and federated learning for privacy-preserved data sharing in industrial IoT. IEEE Transactions on Industrial Informatics. 2020;**16**(6):4177-4186

[128] Oruganti SK. Wireless power-data transmission for industrial Internet of Things: Simulations and experiments. IEEE Access. 2020;**8**:187965-187974

[129] Qadri YA. The future of healthcare Internet of Things: A survey of emerging technologies. IEEE Communications Surveys & Tutorials. 2020;**22**(2):1121-1167

[130] Radoglou-Grammatikis P. A compilation of UAV applications for precision agriculture. Computer Networks. 2020;**172**:107148

[131] Shafiq M. CorrAUC: A malicious bot-IoT traffic detection method in IoT network using machine-learning techniques. IEEE Internet of Things Journal. 2020;**8**(5):3242-3254

[132] Tuli S. HealthFog: An ensemble deep learning based smart healthcare system for automatic diagnosis of heart diseases in integrated IoT and fog computing environments. Future Generation Computer Systems. 2020;**104**:187-200

[133] Al-Qerem A. IoT transaction processing through cooperative concurrency control on fog–cloud computing environment. Soft Computing. 2020;**24**(8):5695-5711

[134] Almiani M. Deep recurrent neural network for IoT intrusion detection system. Simulation Modelling Practice and Theory. 2020;**101**:102031

[135] Cheng JC. Data-driven predictive maintenance planning framework for MEP components based on BIM and IoT using machine learning algorithms. Automation in Construction. 2020;**112**:103087

[136] Cook AA. Anomaly detection for IoT time-series data: A survey. IEEE Internet of Things Journal. 2020;**7**(7):6481-6494

[137] Elhoseny M. Hybrid optimization with cryptography encryption for medical image security in Internet of Things. Neural Computing and Applications. 2020;**32**(15):10979-10993

[138] Gao H. Context-aware QoS prediction with neural collaborative filtering for Internet-of-Things services.

IEEE Internet of Things Journal. 2020;7(5):4532-4542

[139] Hui H. A novel secure data transmission scheme in industrial Internet of Things. China Communications. 2020;**17**(1):73-88

[140] Islam M. Development of smart healthcare monitoring system in IoT environment. SN computer Science. 2020;**1**(3):1-11

[141] Liang W. Deep reinforcement learning for resource protection and real-time detection in IoT environment. IEEE Internet of Things Journal. 2020;7(7):6392-6401

[142] Liao H. Learning-based context-aware resource allocation for edge-computing-empowered industrial IoT. IEEE Int. of Things Journal. 2020;7(5):4260-4277

[143] Lv Z. Interaction of edge-cloud computing based on SDN and NFV for next generation IoT. IEEE Internet of Things Journal. 2020;7(7):5706-5712

[144] Paiola M. Internet of Things technologies, digital servitization and business model innovation in BtoB manufacturing firms. Industrial Marketing Management. 2020;**89**:245-264

[145] Samir M. UAV trajectory planning for data collection from time-constrained IoT devices. IEEE Transactions on Wireless Communications. 2020;**19**(1):34-46

[146] Singh SK. Block IoT intelligence: A blockchain-enabled intelligent IoT architecture with artificial intelligence. Future Generation Computer Systems. 2020a;**110**:721-743

[147] Singh S. Convergence of blockchain and artificial intelligence in IoT network for the sustainable smart city. Sustainable Cities and Society. 2020b;**63**:102364

[148] Tewari A. Security, privacy and trust of different layers in Internet-of-Things (IoTs) framework. Future Generation Computer Systems. 2020;**108**:909-920

[149] Wazid M. LAM-CIoT: Lightweight authentication mechanism in cloud-based IoT environment. Journal of Network and Computer Applications. 2020;**150**:102496

[150] Xu X. BeCome: Blockchain-enabled computation offloading for IoT in mobile edge computing. IEEE Transactions on Industrial Informatics. 2020;**16**(6):4187-4195

[151] Abdel-Basset M. Energy-aware marine predators algorithm for task scheduling in IoT-based fog computing applications. IEEE Transactions on Industrial Informatics. 2021;**17**(7):5068-5076

[152] Abdulkareem KH. Realizing an effective COVID-19 diagnosis system based on machine learning and IOT in smart hospital environment. IEEE Internet of Things Journal. 2021;**8**(21):15919-15928

[153] Alazab M. Multi-objective cluster head selection using fitness averaged rider optimization algorithm for IoT networks in smart cities. Sustainable Energy Technologies and Assessments. 2021;**43**:100973

[154] Ayvaz S. Predictive maintenance system for production lines in manufacturing: A machine learning approach using IoT data in real-time. Expert Systems with Applications. 2021;**173**:114598

[155] Bera B. Designing blockchain-based access control protocol in iot-enabled

smart-grid system. IEEE Internet of Things Journal. 2021;**8**(7):5744-5761

[156] Cai X. A multicloud-model-based many-objective intelligent algorithm for efficient task scheduling in Internet of Things. IEEE Internet of Things Journal. 2021;**8**(12):9645

[157] Chen Y. Energy efficient dynamic offloading in mobile edge computing for Internet of Things. IEEE Transactions on Cloud Computing. 2021a;**9**(3):1050-1060

[158] Chen Y. Deep reinforcement learning-based dynamic resource management for mobile edge computing in industrial Internet of Things. IEEE Transactions on Industrial Informatics. 2021b;**17**(7):4925-4934

[159] Elayan H. Digital twin for intelligent context-aware IoT healthcare systems. IEEE Internet of Things Journal. 2021;**8**(23):16749-16757

[160] Elsisi M. Deep learning-based industry 4.0 and Internet of Things towards effective energy management for smart buildings. Sensors. 2021;**21**(4):1038

[161] Fatorachian H. Impact of Industry 4.0 on supply chain performance. Production Planning & Control. 2021;**32**(1):63-81

[162] Gheisari M. OBPP: An ontology-based framework for privacy-preserving in IoT-based smart city. Future Generation Computer Systems. 2021;**123**:1-13

[163] Goudarzi M. An application placement technique for concurrent IoT applications in edge and fog computing environments. IEEE Transactions on Mobile Computing. 2021;**20**(4):1298-1311

[164] Hu H. AoI-minimal trajectory planning and data collection in UAV-assisted wireless powered IoT networks. IEEE Internet of Things Journal. 2021;**8**(2):1211-1223

[165] Iwendi C. A metaheuristic optimization approach for energy efficiency in the IoT networks. Software: Practice and Experience. 2021;**51**(12):2558-2571

[166] Javaid M. Internet of Things (IoT) enabled healthcare helps to take the challenges of COVID-19 Pandemic. Journal of Oral Biology and Craniofacial Research. 2021;**11**(2):209

[167] Khalaf OI. Optimized dynamic storage of data (ODSD) in IoT based on blockchain for wireless sensor networks. Peer-to-Peer Networking and Applications. 2021;**14**(5):2858-2873

[168] Lin Z. Supporting IoT with rate-splitting multiple access in satellite and aerial-integrated networks. IEEE Internet of Things Journal. 2021a;**8**(14):11123-11134

[169] Lin JCW. Privacy-preserving multiobjective sanitization model in 6G IoT environments. IEEE Internet of Things Journal. 2021b;**8**(7):5340-5349

[170] Liu X. QoS-guarantee resource allocation for multibeam satellite industrial Internet of Things with NOMA. IEEE Transactions on Industrial Informatics. 2021a;**17**(3):2052-2061

[171] Liu S. Fuzzy-aided solution for out-of-view challenge in visual tracking under IoT-assisted complex environment. Neural Computing and Applications. 2021b;**33**(4):1055-1065

[172] Liu RW. Data-driven trajectory quality improvement for promoting intelligent vessel traffic services in 6G-enabled maritime IoT systems. IEEE Internet of Things Journal. 2021c;**8**(7):5374-5385

[173] Liu C. Cell-free satellite-UAV networks for 6G wide-area Internet of Things. IEEE Journal on Selected Areas in Communications. 2021d;**39**(4):1116-1131

[174] Luna-Perejón F. IoT garment for remote elderly care network. Biomedical Signal Processing and Control. 2021;**69**:102848

[175] Lv Z. Trustworthiness in industrial IoT systems based on artificial intelligence. IEEE Transactions on Industrial Informatics. 2021a;**17**(2):1496-1504

[176] Lv Z. Big data analytics for 6G-enabled massive Internet of Things. IEEE Internet of Things Journal. 2021b;**8**(7):5350-5359

[177] Neelakandan S. IoT-based traffic prediction and traffic signal control system for smart city. Soft Computing. 2021;**25**(18):12241-12248

[178] Nagarajan VD. Artificial intelligence in the diagnosis and management of arrhythmias. European Heart Journal. 2021;**42**(38):3904-3916

[179] Poluru RK. An improved fruit fly optimization (IFFOA) based cluster head selection algorithm for Internet of Things. International Journal of Computers and Applications. 2021;**43**(7):623

[180] Qiu H. Adversarial attacks against network intrusion detection in iot systems. IEEE Internet of Things Journal. 2021;**8**(13):10327-10335

[181] Ratta P. Application of blockchain and Internet of Things in healthcare and medical sector: Applications, challenges, and future perspectives. Journal of Food Quality. 2021;**2021**

[182] Shi X. Large-area display textiles integrated with functional systems. Nature. 2021;**591**(7849):240-245

[183] Stergiou CL. IoT-based big data secure management in the fog over a 6G wireless network. IEEE Internet of Things Journal. 2021;**8**(7):5164-5171

[184] Wang X. A multi-objective home energy management system based on Internet of Things and optimization algorithms. Journal of Building Engineering. 2021;**33**:101603

[185] Xiong J. An AI-enabled three-party game framework for guaranteed data privacy in mobile edge crowdsensing of IoT. IEEE Transactions on Industrial Informatics. 2021;**17**(2):922-933

[186] Zhang WZ. Secure and optimized load balancing for multitier IoT and edge-cloud computing systems. IEEE Internet of Things Journal. 2021;**8**(10):8119-8132

[187] Zhao X. Nanogenerators for smart cities in the era of 5G and Internet of Things. Joule. 2021;**5**(6):1391-1431

[188] Zhu M. Making use of nanoenergy from human–nanogenerator and self-powered sensor enabled sustainable wireless iot sensory systems. Nano Today. 2021;**36**:101016

Chapter 2

Future Internet of Things: Connecting the Unconnected World and Things Based on 5/6G Networks and Embedded Technologies

Seifeddine Messaoud, Rim Amdouni, Adnen Albouchi, Mohamed Ali Hajjaji, Abdellatif Mtibaa and Mohamed Atri

Abstract

Undeniably, the Internet of Things (IoT) ecosystem keeps on advancing at a fast speed, far above all predictions for growth and ubiquity. From sensor to cloud, this massive network continues to break technical limits in a variety of ways, and wireless sensor nodes are likely to become more prevalent as the number of Internet of Things devices increases into the trillions to connect the world and unconnected objects. However, their future in the IoT ecosystem remains uncertain, as various difficulties as with device connectivity, edge artificial intelligence (AI), security and privacy concerns, increased energy demands, the right technologies to use, and continue to attract opposite forces. This chapter provides a brief, forward-looking overview of recent trends, difficulties, and cutting-edge solutions for low-end IoT devices that use reconfigurable computing technologies like FPGA SoC and next-generation 5/6G networks. Tomorrow's IoT devices will play a critical role. At the end of this chapter, an edge FPGA SoC computing-based IoT application is proposed, to be a novel edge computing for IoT solution with low power consumption and accelerated processing capability in data exchange.

Keywords: IoT, challenges, AI, 5/6G networks, FPGA SoC

1. Introduction

Lately, the whole field of networks has undergone a significant technical revolution. Network automation is a trendy issue that has been discussed for a long time. IoT technology complements it, which paves the way for the provision of this aspect. The Internet of Things [1] is a cross-device environment created by gadgets that focus on three key tasks: data transmission, data reception, and data processing. Initially, local physical devices connected to the Internet for real-time data analysis were considered the IoT network. The size of IoT has grown over time, from local workstations to industrial IoT frameworks [2]. IoT research describes the proliferation of IoT in healthcare [3], industry setup [4], business analytics, education, area networks, and more. Therefore has the associated risks are due to the expected increase in IoT devices in a diverse environment.

The Internet of Things is one of the most critical and revolutionary trends of the twenty-first century. The Internet of Things (IoT) is a global network of billions of interconnected "things" that can detect, act, and communicate with one another and/or the Internet [1, 2]. Current forecasts exceed initial forecasts for IoT growth: While Gartner predicts 14.2 billion interconnected things in 2019 (which might rise to 25 billion by 2021 [3]), Arm predicts one trillion additional devices will be manufactured between 2017 and 2035 [4]. This tendency is generating exponential increase in the number of chances for businesses and service providers by affecting all sectors of technology, allowing today's organizations, large and small, to gather data on basically everything, from anywhere, at any time. The rise of IoT would be inextricably connected to the wireless trend, which began with Radio Frequency Identification (RFID), also benefiting from the continued development of other conventional technologies naming Wi-Fi, Bluetooth and devices based on IEEE 802.15.4., extensively utilized in traditional wireless sensors [5]. Those kind of systems are typically ad hoc wireless networks made up of a massive number of nodes, i.e. nodes, with limited resources, which work unitedly to reach a common goal (e.g., environmental monitoring and intelligent traffic control, industrial, surveillance systems, etc.) which is capable of transforming physical phenomena into digital data and move them to the Internet.

In the last few years, Motes have been used in a variety of sectors feedback systems, process control, counting monitoring, automotive and automation. Nevertheless, while developing these devices with limited resources, the requirements of small size, weight, low power consumption, and low cost (SWaP-C) are always sought. Physical constraints would continue and be increased by the demands of recent trends as technologies around the IoT edge expand rapidly and boost their potential, namely: (i) Data transfer over the Internet to specific online services in a standardized manner is enabled by connectivity and subsequent interoperability [6–8]; (ii) the need for higher intelligence at the network's edge, allowing systems to make choices faster while consuming less energy [9, 10]; (iii) devices developed for security, mitigating risks from a large number of massive attack surfaces present in the IoT network [11, 12]; and (iv) new energy-saving techniques, allowing autonomous and durable devices [13, 14].

Recent advances in reconfigurable computer technology, specifically Field Programmable Gate Arrays (FPGAs), continue to support the IoT field [3]. Even with low-end IoT endpoints, programmable hardware may give performance advancement, flexibility, scalability [15], hardware-enhanced security, and improved power ratios, making it a suitable choice to handle a wide range of difficulties. Using modern FPGAs in IoT allows for a combination of scalable and flexible resources that are aligned with the SWaP-C premises while also allowing the technology to migrate from the cloud to the edge.

This forward-looking chapter presents a concise and forward-looking assessment of the usage of reconfigurable technology on upcoming low-end IoT motes. This chapter is organized in six section. Section refsec1 focuses into the key trends and issues confronting current low-end IoT devices. The section provides a full review and up-to-date explanation of the application of reconfigurable computing technology to solve such trends and difficulties, as well as a comparative examination of current FPGA SoC-based low-end IoT motes. 2. Section 3 presents the connectivity evolution beyond the 5G revolution. In section 4 we present a real QoS-QoR aware CNN FPGA accelerator co-design approach for future IoT word. Finally, we conclude this chapter in Section 5.

2. IoT edge: trends and challenges

There are four primary trends and problems in the design of IoT devices at the network's edge currently: The presence of high levels of attack vectors and security vulnerabilities necessitates the consideration of scalable security primitives early in the process, and there is a growing trend to deploy intelligence at the edge as data collection increases and even more, meaningful decisions are required. There is constant compression of an already low power envelope due to device design.

2.1 Basis for connectivity and interoperability

Myriads of smart devices might now be linked to the internet as IoT becomes more prevalent. A genuinely standard and lightweight communication stack is necessary to provide connection and compatibility among all existing heterogeneous wireless technologies. A variety of wireless technologies have already been used, causing huge communication heterogeneity and interoperability problems when developing linked IoT devices [16, 17]. The IoT infrastructure's variability makes standardization exceedingly challenging. With the presence of many strong competitors competing for the market dominance, "wars of standards" are unavoidable. In addition, no single technology is capable to provide a single solution that fully and simultaneously meets all the requirements of the IoT network, including power consumption, endpoint cost, bandwidth, connection density, latency, quality of service, operational expenses, and range. Normalization, on the other hand, is critical because it lowers barriers and promotes interoperability across different vendors and devices, permitting new goods and services to coexist with long-standing support. Guideline would be critical in the development and diffusion of IoT, since any communication stack must use methodical algorithms and lightweight protocols to save processing power and save energy [18].

The Internet, as known, links billions of devices using Internet Protocol (IP), specifically IP version 4 (IPv4) [8]. Nevertheless, because of the underlying 32-bit addressing method, IPv4 had major scaling issues, that were solved with the development of IPv6. This edition includes a distinctive 128-bit address for every connected device, as well as an updated protocol architecture to support a wide range of IoT-based heterogeneous devices [19]. concerns, numerous standards bodies, including the Institute of the Internet Engineering Task Force (IETF) and Electrical and Electronics Engineers Standards Association (IEEE-SA), have outlined a foundation for developing communication protocols and wireless technologies that will be implemented using the IoT market [6]. The IEEE 802.x family of standards was one of these organizations' most popular achievements. The IEEE 802.15.4 standard, that specifies a short-range radio frequency transmission protocol for low-power lossy (LLN) networks, low-power, low-rate, has aided in the seamless transition of wireless systems, existing wireless sensors to Internet-connected low-end devices [8]. In addition to its physical (PHY) and medium access control (MAC) layers, additional protocols (e.g., ZigBee, Thread, ISA100.11a, WirelessHART, and so on) have arisen, expanding the heterogeneity of the IoT domain. For the present, the IETF IPv6 over Low Power WPAN (6LoWPAN) working group committed to the definition of the 6LoWPAN adaption layer, that allows IPv6 datagrams to be sent across IEEE 802.15.4 networks. The collaboration of IEEE 802.15.4 compliant radios with the 6LoWPAN protocol allows for easy integration of limited devices with the Internet, which seems to be an important factor in interoperability and communication between low-end IP devices [6, 8, 18, 19].

2.2 Edge intelligence

Massive volumes of data are created, processed, communicated, saved, and analyzed when connection and internet technologies are implemented on in-vehicle devices and the IoT. According to the International Data Corporation (IDC), by 2025, the volume of data generated globally would be predominantly from the edge and would exceed 163 zettabytes (over 1000 billion gigabytes), a tenfold increase over data produced in 2016 [4, 9]. This ideal change would force designers, engineers and technology providers to reconsider how they construct new hardware solutions that go beyond the norm and cope with artificial intelligence (AI) workloads at the edge. Cloud service enterpriser have been at the front line of introducing AI to develop and improve their workloads and services over the last decade. Cloud services will be essential for the next generation of smart industries, smart cities, and smart households. Nonetheless, decreased latency requirements, growing privacy concerns, communication bandwidth constraints, and restricted power budgets have fueled the deployment of intelligence at the edge [10, 20]. Cloud-based decisions should be avoided in safety-critical applications such as autonomous driving since the time it takes to conduct a query/decision might compromise the vehicle's safety, for example, collision avoidance. As a result, local and real-time choices have to take precedence. On top of that, with the end of Moore's Law, we could never rely on the rising and heavy processing power of cloud core technologies to handle the quantity of data created by next-generation IoT systems [21]. Cloud services will be critical for doing high-level analytics, yet AI deployment at the edge is also increasingly critical. Deploying and utilizing intelligence at the edge has inherent dangers as well as a set of needs, both in terms of security and SWaP-C. In terms of security, the increasing complexity of the edge exponentially widens the security flaws, bringing up new attack routes in an infrastructure that is already striving to give increased protection. Advanced computing techniques, such as machine learning, can greatly increase the processing capabilities of wireless sensor nodes while also lowering total network power consumption through decreased wireless transmissions [22]. Pushing these tasks (data analysis and decision inference) as far as feasible would eventually optimize resource efficiency and responsiveness, leading to more autonomous and intelligent systems [23].

2.3 Security

Security in the Internet of Things era is not voluntary, and it should be a fundamental layout priority from the start and throughout the device's lifecycle. As IoT grows deep inside important enterprise infrastructures, the value of the assets contained within such devices rises, making them attractive targets for attackers and hackers. Therefore, ignoring device safety as an initial design issue may endanger the whole supply chain, resulting in revenue and brand reputation risks, as well as grave life-threatening circumstances in some cases. The success of the Internet's next phase is highly reliant on the inherent trust and security of billions of linked heterogeneous network devices [6, 11, 12].

A flexible, multi-layered strategy capable of providing end-to-end security from device to cloud and everything in between is necessary to deliver a security architecture solution that completely covers an IoT platform. While the majority of the initial architectural proposals involves a three-layer design (perception, network, and application layers) [11], a common dominating choice has yet to be determined. In subsequent versions, more abstraction has been incorporated, culminating in a five-layer structure (service management, object abstraction, application layer, objects and business layer [24]. Each layer's technologies are distinct, with their own set of goals, needs, constraints, and tradeoffs. Nonetheless, the IoT's diverse set of security challenges and vulnerabilities has an inherent impact on all layers of the architecture.

Info concerning the IoT architecture was transmitted across all levels and entities, i.e., users, service providers, and devices, to ensure full compatibility between services and devices. This, however, considerably expands the entire attack surface. The four primary categories of attacks include hardware-based attacks (e.g., changing techniques or channel violent attacks), communication attacks (e.g., weak random number generators, man-in-the-middle), life cycle attacks (e.g., degradation code, oversupply at the factory) and software attacks (e.g., return oriented programming approaches, malware). Countermeasures must be implemented for each form of attack because a single weakness may split the entire device and span the whole network. A list of technologies and mitigation methods can be chosen based on the offered assets of an IoT-based product to fulfill the essential security standards that must be enforced. Meeting these standards is essential in establishing a reliable and secure IoT infrastructure that provides rigorous guarantees on security primitives. Among these security primitives are the following:

- Authentication is an essential aid in ensuring the security of communications between different parties [25]. The first barrier of protection against intrusion is access control management. These mechanisms are essential in order to identify and classify objects and manage their identities, establish a mutual trust relations between various objects, users, or systems by verifying and distinguishes their identities, and grant, deny, or limit entities' access to data, resources, or applications (i.e., authorization) [26].

- Resource availability is one of the fundamentals of IoT security which may be maintained through strict hardware maintenance and safe software/hardware resources. Additional security measures, for example software firewalls and intrusion detection systems, could be deployed to prevent malicious behaviors like denial of service (DoS) attacks.

- Information authenticity is linked to the source of the data [27]. End-to-end security methods are required to guarantee that data is coming from valid sources. Globally notable identifiers and hierarchical identification methods are fundamental to assuring IoT authenticity [28].

- Integrity is about maintaining the consistency of data, ensuring that unauthorized entities, or even unidentified causes cannot modify it undetected [29]. Cryptography is commonly used to verify data integrity.

- Privacy aims to prevent important information from reaching the hands of unauthorized people or devices. This is usually accomplished by establishing several degrees of access for the wanted asset (user/password), biometric verification, two-factor authentication, robust data encryption technologies, security tokens, and other methods.

Since no security primitive by itself provides a standardized solution, it is essential to take a relevant layered approach to give the complex foundation to copiously defend the entire IoT device architecture, infrastructure, commonly known as defense in depth [30]. **Figure 1** shows a high-level overview of the various types of security solutions [31]. All of the layers lead to strengthen the safety of the IoT system, and every one addresses a distinct security issue.

- The Foundation Functions layer provides core modules that service the layers above it, such as cryptographic algorithms/engines-backed true random number generator (TRNG) modules [31]. The following cryptographic schemes stand out from the rest: (i) the Advanced Encryption Standard (AES) symmetric key

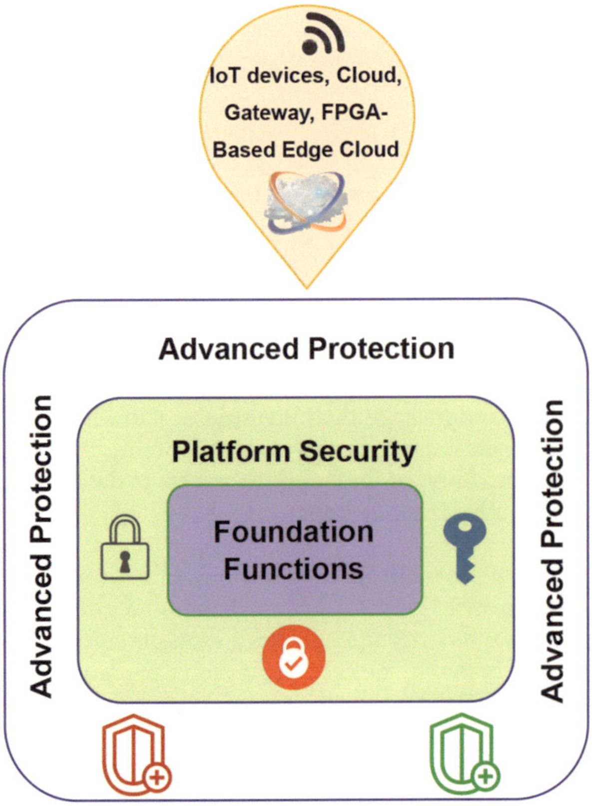

Figure 1.
A layered overview of the main security technologies used in IoT (adapted from [31]).

protocol for mass information encryption, (ii) the Secure Hash Algorithm (SHA) cryptographic functions, and (iii) the Elliptical Curve Cryptography (ECC) or RSA asymmetric key algorithms for authentication and secure session key transactions. This layer offers a system that provides a special device identification, that is silicon bound, to enable various cryptographic algorithms (root key) [31]. A root key is usually held in a single-use programmable memory, which is configurated during platform manufacturing, or in physically unclonable function (PUF) mechanisms. It gives a strong method for encrypting, additional keys and data.

- The system security layer is concerned with a system-wide approach to platform security, integrating device and memory access management. Memory protection units (MPUs) are commonly utilized for this purpose. ARM, on the other hand, has lately moved its TrustZone technology, which was previously limited to its microprocessors (Cortex-A), to the level of microcontrollers (Cortex-M). The latter helps to divide applications that handle sensitive peripherals or memory sections of the operating system as well as other hardware modules on the platform. Arm TrustZone-M advocates hardware as the first root of trust and allows any system resource (e.g., CPU, memory, and peripherals) to be trusted. The security layer of the platform is also in charge of verifying the integrity and authenticity of the software executing in the IoT device. Secure boot is the key technology in this regard [31].

- The advanced protection layer contains a collection of technologies that defend physical tampering threats that might compromise the system's integrity, availability or confidentiality. As a result, this layer includes technologies to stop: unauthorized access to the IP code, data, or keys (confidentiality), unauthorized changes to the code, data, or keys stocked in the apparatus for trying to take control of the system (integrity), and methods for interrupting the system's normal operation, rendering it not available or operating in safe mode (availability). Also physical sabotage attacks, whether or not they involve physical attacks, can be classed as invasive or non-invasive. Infiltration or damage to the device's packaging, respectively [32]. While detecting invasive attacks is simple with an on/off switch connected to a treatment system's GPIO pins, detecting non-invasive attacks is significantly more costly.

Non-invasive attacks are often classified into three categories: side-channel attacks, fault injection attacks, and software attacks [32]. Side-channel attacks concentrate on monitoring the system's behavior in terms of time (temporal attack), electromagnetism, and power consumption, simple power analysis (SPA), and differential power analysis (DPA), while 'it executes secure operations (e.g. cryptography) to extract the keys. The most effective technique to prevent synchronization attacks is to ensure that all operations inside a security function spend the same amount of time. Intel has solved this issue by developing a fully dedicated Advanced Encryption Standard (AES) instruction set that operates data-independent. Kocher [33] presented a platform-independent method for updating the secret key for each executive session of a cryptographic scheme, causing the synchronization patterns. Rambus [34] suggested a set of software libraries and hardware cores that are immune to secondary channel attacks such as temporal, electromagnetic, SPA, and DPA attacks. In fact, their methods are based on strategies that reduce the signal-to-noise ratio on side channels and introduce randomization into cryptographic operations. They even implement protocol-level countermeasures, changing cryptographic protocols to include key update methods.

2.4 Energy awareness

Recent technical advancements in the information and communication technologies (ICT) industry have come at a cost, which is now associated with a 2% increase in the average carbon footprint. Nonetheless, because of the increasing of ICT scenarios and their requirements (including a massive and promising IoT ecosystem), it is predicted that by 2020, ICT improvement would be in the range of 6–8% [13]. The rapid spread of IoT technologies and their broad acceptance will require further sensory, communication, and performance add-ons, putting even more pressure on these devices' energy budgets. On the other hand, while IoT infrastructure will boost carbon footprint over the next few years, it also has the potential to be explored to minimize the environmental footprint of several major sectors of society: habitat monitoring, energy, smart cities and transportation systems (e.g., smart grid, smart traffic jam, etc.).A smart grid anchored by IoT nodes, for example, may improve total energy consumption. From a macro and "green" standpoint, IoT devices require a more efficient and sustainable use of resources, with the problem of energy consumption at the heart of any IoT system's design and development [13, 14].

IoT devices should use minimal power as possible. Because these devices require continuoual technique indefinitely, stable and reliable power sources are important enablers: repair and replacement of the battery or device are not cost-effective methods. Recent advancements in energy harvesting systems provide fundamental approaches for increasing battery life, mobility, and range [35, 36]. Furthermore, system designers must rely on existing and next-generation power management strategies (e.g., low-leak processing technologies, low-power flash memory and non-volatile memory technologies, low-power clock and operational diagrams, protocols) to minimize the total energy budget. The effective and sustainable use of power resources is critical since energy consumption determines the life of a particular battery capacity [37], which necessitates the implementation of a set of control methods and intelligent energy management. As a general rule, Motes often function cyclically, periodically alternating phases of active and low power operation to reduce their average power consumption and hence lengthen their longevity [36]. When a device is in active operation, it often demands wireless communications, that is commonly needs the most power state of a node. In brief, as the IoT's backbone, wireless sensors would address rising energy demands and problems by introducing new energy- functional primitives.

3. Roles of reconfigurable platforms

Over the past years, the semiconductor enterprise has consistently reduced the size of its devices while increasing their power and efficiency. Moore's Law drove down the cost per transistor dramatically each time the total number of transistors created was duplicated (approximately 45%) [36]. The pace with the fast demand for quicker and smaller goods has driven this technology to its limitations, making it increasingly hard to rise the density of transistors on a chip also its operating frequency, that appears to be nearly saturated [21]. This Moore's Law deceleration raises several challenges for system designers, who expected better performance-to-energy ratios from each new generation of devices. This technical deadlock has prepared the way for the introduction of reprogramed platforms (i.e., FPGA-based platforms) as a novel hardware method to addressing these difficulties across a wide range of on-board use areas [5, 15].

Figure 2 illustrates the various software and hardware architectures that are now available on the market and commonly employed in the creation of embedded systems [21]. Microcontrollers (MCUs) provide the most flexibility, whereas ASICs (application-specific integrated circuits) offer the maximum performance. FPGA-based solutions, on the other hand, could offer the best of both worlds by providing high crippled processing abilities, leading in greater performance raise than MCUs and the ability to be reprogrammed at any moment. In comparison to ASICs, over time execution via partial or dynamic reconfiguration techniques provides greater flexibility. However, because MCUs are software devices, they offer best flexibility, making them useful in basic, low-cost embedded systems [38].

FPGA manufacturers have begun to integrate embedded processors (soft or hard) into their gadgets in recent years, leading to so-called Field Programmable Chip Systems (FPGA SoC), which have emerged as the world's greatest option for balancing flexibility with efficient computing power. FPGA SoCs have progressed from a single or dual-core processor-only platform to a far more powerful platform with graphics processing units (GPUs), real-time processors, multi-core processors, real-time processors and specialized hardware blocks such as digital signal processors (DSPs) and video compression components. With this varied array of resources ranging from systems that are efficient intended at high-end applications to a better resource-constrained platform, these heterogeneous reprogrammed technologies are better technical options for dealing with the ever-increasing diversity of Low-end IoT applications. Nevertheless, depending on the target context and uses situation, each IoT deployment may use a distinct network data and transmission architecture, technology and design processes that are widely used, based on the general requirements imposed by the environment's natural evolution the IoT ecosystem.

3.1 Connectivity and interoperability

Hardware-assisted technologies that can speed widely recognized protocols and standards at the network edge are steadily resolving connectivity and interoperability problems in reconfigurable systems. For example, some data privacy-related

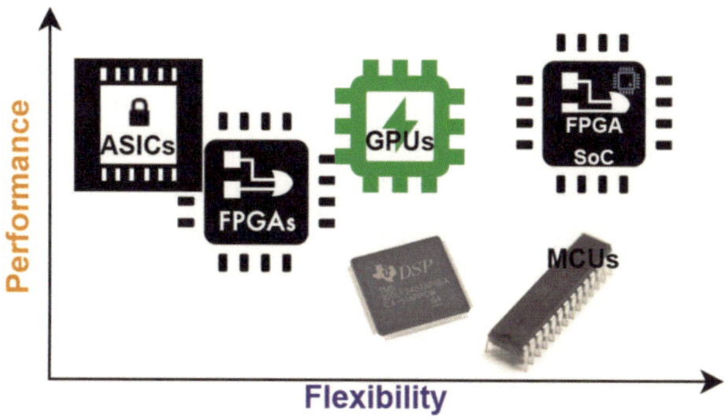

Figure 2.
Performance versus flexibility of different processing platforms.

communication operations (e.g., authentication, data encryption/decryption) take quite a long time and cost a lot of power. Offloading such activities to hardware (e.g., cryptographic protocols and algorithms) can result in improved performance-to-power tradeoffs. Gomes et al. [39] suggested a 6LoWPAN accelerator that can analyze and filter packets received by a radio transceiver. When compared to software filtering, the findings demonstrated a nearly 13.24% reduction in performance overhead. In addition to speeding up these computing operations, reconfigurable systems can help to reduce the obsolescence of cryptographic primitives through dynamic partial reconfiguration (DPR) [40]. Furthermore, some IoT-based applications have consistently employed FPGAs for networking reasons and obtained promising results in recent years, fostering the growth of different solutions in the industry. In [38], Andina et al. discussed many research that demonstrate the benefits of employing FPGAs to tackle connectivity challenges on IoT systems.

The growth of software-defined radio systems has coincided with the evolution of radio communications (SDR). An SDR is a radio communication system in which standard FPGA hardware components (e.g., mixers, filters, amplifiers, modulators/demodulators, detectors) are integrated in software. Indeed, this method facilitates the generation of smart communication strategies with great usefulness in a variety of sectors (e.g., mobile phones or military applications), where protocols and radio settings (e.g., new modulation designs, filters) may be modified in real time. The benefits of reprogrammable platforms paired with the SDR paradigm give up a new pair of possibilities in which new hardware modules (specified in software but speeded in hardware) may be developed and dynamically installed on reconfigurable systems using DPR [41, 42].

3.2 Intelligence

The current trend to solve the problems of excess information created at the edge and latencies engendered by its transmission through the network has given rise to the concept of edge computing, in which the edge node uses AI, specifically deep learning methodologies, to properly accomplish data analysis at the source. Because of their inherent parallel compilation ability and performance per watt benefits, FPGA-based platforms are well suited to address AI needs in this situation. By offering hardware-accelerated inference techniques, these systems can fulfill the strict effectiveness and power restrictions of edge devices.

The latest generation low density FPGAs, such as the Xilinx 7000 family, can speed neural networks in the 1 W to 1 mW region. Each FPGA series has a convolutional neural network (CNN) accelerator that may be configured for accuracy or power consumption [20]. In comparison to previous platforms, Intel's new FPGA SoC combine DSP blocks with unique floating point capabilities into the FPGA fabric, considerably reducing logic resources consumption and improving overall performance [9]. To be fair, the high computational storage and power capacities requirement of classic neural network designs continue to confront even the newest FPGA-based platforms.

While the market has recently lauded FPGAs' capabilities for AI acceleration, academia has as well thoroughly researched this subject, presenting many accelerators and demonstrating interesting results. In [43] the authors developed a low-precision CNN accelerator that delivered about the precision of a standard CNN while outperforming other tasks by up to 6 times. The authors of [43] developed a CNN-based image classifier accelerated on a high-end FPGA SoC that investigates both the integrated hardcore and the FPGA fabric holistically. When compared to standard hardware platforms, the solution achieves outstanding performance/power consumption ratios (e.g., CPU, GPU). Other similar papers offer techniques based on hardware accelerated AI as well. In the works cited in [44], the results presented make it possible to accelerate the performance of 4x compared to other solutions while reducing energy consumption. Although the Deep Learning Accelerator written in OpenCL is capable of accelerating AlexNet up to 10 times quicker than the different leading edge approaches. From another point of view, the authors of [45] have proposed a series of efficient design techniques (p. To meet the limitations of devices with limited resources. A common element in all these works is that solutions based on FPGA strike a true balance between compute performance and energy efficiency. Furthermore, these platforms are the only option capable of continuously adapting to the high speed of development in AI frameworks, both in terms of algorithm implementation and performance/power needs for future generation workloads.

3.3 Security

IoT system security is a critical necessity. The attack vector spectrum is expanding, and IoT system developers want solid and very secure countermeasures to efficiently protect the upcoming devices generation. Faced with today's security demands, new reconfigurable systems include a variety of security blocks ranging from core hardware encryption engines.

The basic functions include numerous techniques that support greater security standards, as aa example we mention data encryption/decryption before performing data transmission. The latest generation of FPGAs provide a diverse set of integrated, hardware-accelerated blocks and cryptographic resources (e.g., ECC, AES, SHA, and HMAC). Microsemi's SmartFusion, SmartFusion2, and IGLOO2 devices, for example, provide hardware accelerators for AES-128/256 and SHA-256, that may be utilized for performing design also data security (e.g., to validate the integrity and authenticity of a bit stream). In addition, the advantages of employing FPGA-based cryptographic accelerators have been extensively discussed in the literature; Piedra et al. [46] compared the performance and power consumption of cryptographic primitives in commercial IoT nodes to an FPGA-based cryptographic accelerator. The outcomes shown that the latter technique may significantly improve the execution time of sophisticated cryptographic algorithms, and hence power consumption. Another feature of FPGA-based cryptographic systems is their inherent reconfigurability, that may be used to simply upgrade limited or obsolete cryptographic algorithms and protocols.

A TRNG block is necessary to support cryptographic engines. It generates random cryptographic keys from a statistically independent source of random values. While exhibiting many physical sources of entropy (e.g., clock jitter, thermal noise, shot noise, etc), FPGA-based TRNGs may as well attain ideal high-speed ratios and function as a source of truly random numbers [47]. Newer FPGA-based applications, such as Microsemi's FPGA SoC, are also equipped with a non-deterministic TRNG that is certified to handle cryptographic applications. The protected root keys, which must be uniquely tied to the device, are another important feature of the basic function class [31]. Today's cutting-edge implementation depends on PUF technology, which, due to unavoidable differences in the nanoscale manufacturing process, makes PUFs a viable physical device attribute for generating a peculiar silicon fingerprint [48]. Root keys created from PUFs are fetched from the chip rather than kept on it. PUFs are now present in a wide range of devices, from small sensors and microcontrollers to FPGA-based systems. PUF-based applications in [49] serve as a means for security software on an MCU as well as a basis for authenticating IoT devices in the cloud. On contrast [50], proposes a secure protocol based on PUF to secure a DPR compliant IoT design deployed in FPGA.

The wide system solutions that provide secure primitives at the CPU level provide platform security as well as access control to system components (e.g., FPGA blocks, peripherals, memory). Arm, the dominating architecture in the mobile and in-vehicle categories (with 50 billion devices deployed), launched the most powerful current platform security mechanism, Arm TrustZone, in 2014. Arm TrustZone is a hardware security system that covers both low and high-end Arm CPUs. The later provides a compartmentalized method to security by giving two hardware-reinforced regions of protection: secure and regular worlds. The different worlds are totally separated from hardware and have uneven privileges, preventing insecure software from immediately accessing secure global components. The Trust Zone bit is not contained within the CPU; it extends from the processor to the bus to the hardware's internal circuitry, Zynq-based FPGA SoC are a great example. This technology has been widely employed in academia and business as a significant enabler for the use of Trusted Execution Contexts (TEEs) and to offer strict isolation (security by separation) in critical environments.

3.4 Energy

Users would expect tinier, smarter, and longer-lasting IoT items offered by ultra-low power IT-optimized solutions. FPGAs have lowered power consumption per operation by more than a factor of 1000 since their introduction [51]. These advancements have been driven mostly by process technology and the desire to reach new markets, particularly the consumer sector. Power concerns are now at the forefront of FPGA architecture considerations, and newly FPGA categories are all geared towards low-cost, high-volume applications. The majority of FPGAs are based on SRAM technology, which necessitates extra non-volatile memory to keep their configuration pattern, increasing power consumption. Nonetheless, these platforms have changed significantly over time, and newer devices are more energy effective. For example, Lattice's iCE40 family of FPGAs can operate at 10 mA in active mode and up to 35 μA in standby mode. Due to the uncertain initial state of the SRAM cells, Lattice systems are prone to spikes in starting current (inrush current) like SRAM-based FPGAs. iCE FPGAs, on the other hand, have a maximum inrush current of 1.2 mA, which is a very high efficient number for battery-powered uses.

Flash FPGAs have always fallen behind SRAM-based devices in regards of performance, density, and on-chip IP. Although, new developments in flash technology (e.g., flash memory cell reduction, flash memory integration into advanced logic operations) have dramatically increased these platforms. This technique has a very low static energy consumption as well as minimal inrush and setup power. Microsemi FPGAs investigate flash memory. The IGLOO series, specifically developed for today's portable and energy efficient devices, may deliver standby power consumption rates as low as 2 μW in their FPGA portfolio. Furthermore, Microsemi's FPGA SoC have Flash * Freeze technology, that places the FPGA design in a low-power sleep mode while maintaining the prior state of memory, enabling for quick FPGA shutdown and restart. Sensor arrays, which are invariably turned on and off on a regular basis, might benefit immensely from this feature. Furthermore, a system designer may take use of the extra combination of such technology with other low power modes provided by the integrated hard-core MCU to fulfill the rigorous energy requirements of numerous IoT applications.

To improve energy efficiency, some solutions incorporate a dynamic power management (DPM) module in their reconfigurable hardware, that permits individual resources to be totally turn off in standby or low power mode, as well as a reset function. The voltage and frequency dynamic scales used to govern the digital processing component. The later function is a power management approach that allows the voltage and speed of the MCU to be altered and decreased to lower levels when not in use to reduce power consumption. Furthermore, reconfigurable solutions appear to be an excellent option for heterogeneous grains of low power by studying low consumption operating partners with extremely low static power consumption also employing a DPM system paired with DVFS approaches [52].

3.5 Combination of reconfigurable platforms and IoT Motes

The platforms based on FPGA are quite diverse, extending from compact form dimensions, ultra-low power consumption, and production-priced solutions to fully SoC-enabled platforms with considerable hardware resources to fulfill customer nominations this days. This technology, which has a high level of maturity, is a good option for designing personalized solutions for wireless detecting uses. The authors of [53] have published a complete survey addressing a wide range of hardware devices available for low-end IoT mobiles. By providing numerous solutions based on stand-alone FPGA platforms or heterogeneous designs that integrate an MCU and an FPGA, this paper focuses on the rising focus in researching reprogrammed architectures used in this industry. FPGA-based designs have permit the optimization of numerous components of wireless sensors in aspects of performance and power consumption, while some work has also boosted device security.

Several methods aimed at wireless sensor systems have previously been presented, including PowWow [54], CookiesWSN [37], HaLoMote, and CUTE mote [55]. In these references, the recent state of the art, are well highlighted, on low-end IoT motes which leverage reprogrammable technology on their design, describing their variations from previously recognized CGUs, in addition their most essential qualities and attributes: network accelerators available, radio device used, SoC adopted, MCU design, local security related hardware/software, application specific accelerators, and maturity level. PowWow and CookiesWSN are the first low-end motes implementations that integrate a low-power MCU (TIMSP430) with a tiny low-power Flash FPGA as well as a radio transceiver. The first solution looks into using the FPGA to build low-level network-bound accelerators such forward error correction (FEC) methods. PowWow investigates energy management approaches to manage the digital processing element in order to enhance energy efficiency. While both feature an Elliptical Curve Cryptography Accelerator (ECC), CookiesWSN adds an application-specific Sensor Data Processing Accelerator (SDP) as well as a reprogrammed Kalman filter to reduce noisy samples in the process of data acquisition processing. Despite the major accomplishments of PowWow and CookiesWSN, the utilization of discrete MCU and radio frequency (RF) components resulted in slower communications and worse power efficiency.

Recent alternatives, such as HaLoMote and CUTE mote, have solved some of the previous methods' limitations. HaLoMote, a hardware-accelerated low-power mote aimed at IoT, combines an RF-SoC transceiver (ATmega256RFR2) with a Microsemi

IGLOO M1AGL1000 crawled to speed up massive computation tasks mentioning sensor data aggregation in an SDP. Furthermore, the system offers a DPM accelerator, which enables low power standby modes with extremely low static power consumption, resulting in decreased power consumption. The CUTE mote, on the opposite, is described as a programmable and dependable terminal device that is specifically built for low power IoT applications. The design is implemented on an FPGA SoC (Microsemi martFusion2) platform, which combines an Arm Cortex-M3 hardcore MCU closely linked with a Flash-based FPGA and an externally connected IEEE 802.15.4 radio transceiver. Offloaded hardware accelerators are provided as hardware devices to the MCU and are accessed using a standard on-chip communication protocol, which simplifies design and minimizes access time. The contribution in [55] used a micro-positioning measurement system to evaluate and install their platform. A specific application SDP, a root mean square (RMS) statistical procedure for information evaluation and analysis a Fast Fourier Transform (FFT) method for digital differential signal processing, a finite impulse response (FIR) filter for signal processing, and other signal and image compression techniques have been used. Other relevant contributions in this field [56], despite being at a low maturity level, analyze significant improvements in reprogrammed systems dedicated for FPGA-based wireless sensing uses and conforming to standards, Low-end IoT, where it is still suggested to deploy specific, network, and security related tasks in the FPGA. Although they contribute to a common vision, some contributions are still in the design phase.

Despite variations on multiple levels, all of the prior studies referenced have a common point of view, that we defend through the following chapter: Indeed, in the future of IoT-enabled devices reconfigurable platforms will make a crucial role, where essential problems like as connection and interoperability, cutting-edge AI, hardware and energy efficiency and data security, will surely keep being the top trends and difficulties for future low-end IoT Words.

4. Connected the unconnected world and things: an evolution in connectivity beyond the 5G revolution

The future of the connected world is not only about the latest cutting-edge technologies, such as the constellations of high-speed 5G and low earth orbit satellites. Much will be defined by the advancement and development of current advanced connectivity technologies, such as fiber, low to medium band 5G, 6G, and different other long and short-range solutions. The modern connectivity architecture also includes cloud and edge computing which is accessible with less expensive and more efficient devices and platforms as well as the FPGA SoC (discussed in the above section), as depicted in **Figure 3**. Computing power, storage, and sensors are all getting more robust and reasonable. With the converges of these trends, the connectivity ecosystem will be dominated with more technologies, services, and vendors more than before.

The new and improved networks will enable and complement other critical technologies such as cloud computing and FPGA SoC-based edge computing. These developments, when combined, will allow some of the most data-intensive applications of the future. Cloud computing will keep to serve as a processing backbone for use cases that need a high level of computing power, storage capacity, and complex data analysis capabilities. This computing is required for a variety of

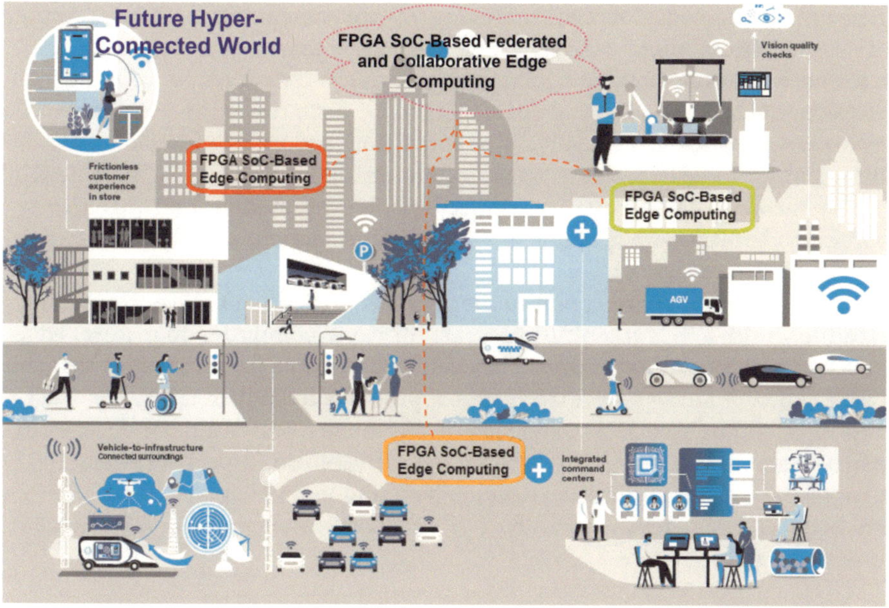

Figure 3.
The future trend of the connected world and things.

tasks ranging from storing films to training artificial intelligence systems. Users' devices may not be able to run the most complex applications without a boost from cloud computing, or they may have to be considerably more expensive. On numerous fronts, FPGA SoC-based edge computing tries to alleviate some of the constraints of cloud computing. Instead of sending data to central cloud servers that may be hundreds or even thousands of miles away from the end user, FPGA edge computing delivers computing power, storage, and networking closer to where data is created or consumed. Actual computing could then take place in smaller-scale data centers on the outskirts of major cities (the metro periphery), at the base of radio access network base stations (the micro-periphery), in wiring closets at end-user premises (the Edge Gateway), or even on the device that generates data itself (the Edge device).

A number of factors are driving the urge to bring processing and storage closer to the end-user. The first is the proliferation of linked devices, particularly as the Internet of Things is implemented in an increasing number of locations. According to a recent IDC [57], prediction, there may be up to 42 billion linked IoT devices by 2025. These technologies are also growing more complicated, progressing from simple smart devices to intelligent linked systems and processes. As the number of increasingly complicated devices grows, so does the volume of data created, which may surpass what a centralized cloud can handle, especially as IoT applications rely more on video processing and ultra-high-definition audio. As a result, there is an increased demand for efficient storage that assures data protection. Another important driver of edge computing growth is the desire for real-time analytics, decision making, and changes. These features are critical for applications such as augmented and virtual reality, linked vehicles, drones, video surveillance, and industrial machinery remote control. This requirement for low latencies reduces transmission time to the cloud.

Also, application development is moving towards new solutions such as container-centric architecture, micro-services architecture, and server-less computing platforms. These solutions provide lightweight, portable alternatives for running applications at the edge, allowing developers to perform testing and maintenance faster and more efficiently. Finally, edge computing addresses a fundamental requirement for industrial operators managing transportation and logistics networks or remote facilities. They may now connect to compute, storage, and analytics resources in contexts with sporadic or restricted connection, as well as in extremely remote locations.

All those different factors point to the upgrading adoption of edge computing around the world. While it took 10–15 years for cloud computing to mature, edge computing is on a faster trajectory. The cloud ushered in a paradigm shift that shifted software and computing power from owned products to delivered services. Edge computing could be seen as an extension of this move towards a more decentralized model. The emphasis today is on defining the architecture (especially emerging industry standards for application development and maintenance, and for interoperability between edge, device, and cloud). Its acceptance may pick in speed once it becomes available.

5. Proposed QoS-QoR aware CNN FPGA accelerator Co-design approach for feature IoT world

5.1 QoS-QoR CNN accelerator for IoT devices

Motivated by the idea and challenges discussed above, we propose a QoS-QoR aware CNN FPGA Accelerator co-design process that includes a hardware-oriented CNN topology and an accelerator design that takes into consideration CNN-specific properties. CNNs and accelerators are created in tandem to find the greatest balance among both QoS and QoR. Targeted QoS, QoR, and hardware resource limitations are inputs to this procedure, while the resulting CNN model and its related accelerator architecture are outputs. The entire process is broken down into three steps:

- Step One: The bundle is created, and the QoS is assessed. We pick CNN components at random out from the pool layer and create bundles (the basic building blocks of the created CNNs) with various layer combinations. Analytical models are used to analyze each of the assemblies in order to capture hardware parameters (e.g., latency, compute and memory needs, resource utilization), allowing a quality of service estimate to be made at the CNN exploration start.

- Step Two: Selection of bundles based on QoR and QoS. To find the far more potential beams, we first analyze each beam's QoR potential by reproducing it n times to create a CNN prototype. For exact results, all CNN prototypes are trained quickly (20 epochs) on the selected dataset. We classify the CNN prototypes with homogeneous QoS to the input targets based on the QoS predicted in step one, and choose the best bundle candidates from each class.

- Step Three: Exploration and exploitation of CNN that is hardware-dependent. We begin exploring CNNs with the first level technique by stacking the selected packet and utilizing stochastic coordinate descent to explore CNNs under provided QoS and QoR restrictions (SCD). The QoS of SCD's CNN outputs is precisely assessed before being sent to SCD for updating the CNN model. To increase QoR, produced CNNs that match QoS criteria are presented in the purpose of training and tuning.

Based on this, we propose an accelerator-based selected CNN design that provides a pipeline architecture for efficient CNN implementation with a maximum resource sharing technique. It contains a foldable structure that uses the same hardware components to calculate CNN sets sequentially, saving resources when targeting tiny IoT devices. To improve QoS, it also uses an unfolded structure for computing operations inside bundles in a pipeline manner. The proposed design can benefit from both recurring and pipeline structures by combining the two levels of design. The acceleration phases, on the other hand, are carried out using Xilinx vivado high level synthesis (HLS).

5.2 Proposed architecture: Acceleration and designing tools

HLS approaches have increased the development quality of FPGA-based hardware design in recent years by enabling FPGAs to be programmed in high-level languages (e.g., C/C++) [58]. Designing an FPGA-based CNN accelerator with high-performances, on the other hand, is far from simple, as it necessitates specific hardware development, repeated hardware/software testing to assure operational accuracy, and efficient design space exploration for advanced throttle settings. We've seen a rising interest in expanding automation frameworks for developing CNN accelerators from a higher level of abstraction, using particular algorithmic descriptions to CNN and top quality predefined hardware models for rapid design and prototyping, in order to increase the effectiveness of accelerator design. However, there are still design issues, as new development patterns in cloud and embedded FPGAs create fundamentally distinct challenges in satisfying the diverse demands of CNN applications. For example, many arrays are frequently employed in the newest versions of cloud FPGAs to double available resources and give better throughput. When accelerator architectures struggle to grow up/down to meet chip size, cross-routing and distributed on-chip memory can simply create timing violations and reduce possible performance. On the other hand, on-board FPGAs combine heterogeneous components (such as a CPU and a GPU) to efficiently handle various aspects of the targeted activities. It is very difficult to fully use on-chip resources and reap the benefits of specific hardware without the need for an extremely flexible task partitioning scheme. Meanwhile, many researchers are experimenting with fast CONV algorithms to see if they can improve the program [59]. While these accelerators deliver superior performance than classic designs, they are constrained by use cases and necessitate more complicated design approaches. As shown in **Figure 4**, the proposed QoS-QoR aware CNN FPGA accelerator co-design is consisted of Zynq Processor that is used in all tasks management, like predictions, GPIO management, and automatically mapping the CNN accelerator with the right parameters. The Axi DMA is used to speed up the data and communication exchange between DDR and CNN accelerator. This co-design aims to test the created CNN accelerators on an edge object detection application.

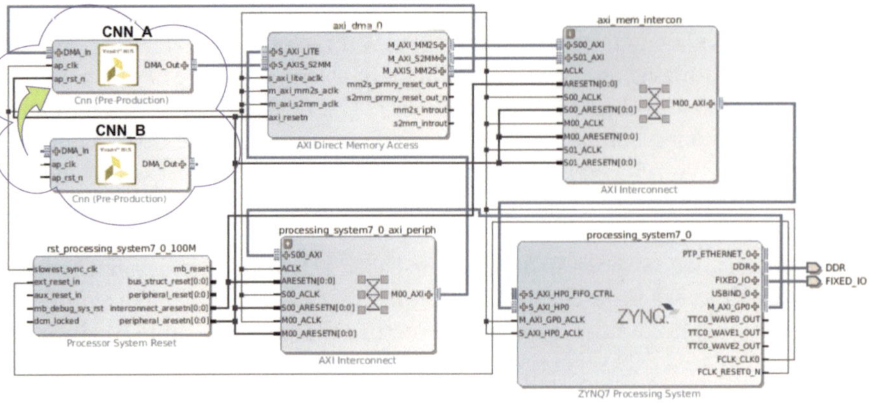

Figure 4.
QoS-QoR aware DNN/CNN FPGA accelerator Co-design.

6. Results and discussion

Considering the work in the literature [60] and in order to test the proposed co-design, we used the same accelerated CNNs (CNN_A, CNN_B) with different layers and configuration summarized in **Table 1**. Th parameters are summarized in different data precision for weights and feature maps. We test then the proposed QoS-QoR aware CNN co-design on an object detection. The suggested co-design schemes finds the most promising CNN topology example for the intended hardware system and application as a bundle containing depth-wise Cnv3 (DW-Cnv3), point-wise Conv1 (PW-Cnv3), and max-pooling layers. Depending on this data, the co-design investigates 3 CNN configurations, each with a distinct normalization strategy, in order to meet the QoR and QoS requirements. **Table 1** shows the different result of the proposed scheme. Using the FPGA Pynq Z1 and the proposed architecture achieved a best results when used different CNNs (CNN_A, CNN_B). According to these results, CNN_A occupies 27% FFs, 78% BRAMs, 84% DSPs, and 76% LUTs with a working frequency of 150 MHz. In addition it reached a 23 FPS with a maximum latency of 44 ms, a maximum power of about 2.6 W and an energy efficiency of about 0.114 J/image. On the other hand, the CNN_B with the configuration of W16 & F8 occupies a 38% FFs, 96% BRAMs, 91% DSPs, and 83% LUTs with a working frequency of 150 MHz. According to this highest hardware cost compared to the first topology, CNN_B cannot surpass the first one considering its energy efficiency factor which is of about 0.16 J/image.

7. Conclusion

This forward-looking chapter provides an outlook on low-end motes in the age of IoT. It illustrates how current reprogrammable platforms are the best choice to adapt to the ever-changing IoT environment after a full assessment of the trends and problems offered by the IoT paradigm to low-end devices. Obviously, the ever-increasing volume of data created by IoT motes, along with the end of Moore's Law, necessitates

CNN Topologies	CNN_A (W16 & F16)	CNN_B (W16 & F8)
Input Layer	Input RGB Image (160*160)	Input RGB Image (160*160)
Layer 1	DW-Cnv3 (3)	DW-Cnv3 (3)
Layer 2	PW-Cnv1 (48)	PW-Cnv1 (48)
Layer 3	Max-Pool (2*2)	Max-Pool (2*2)
Layer 4	DW-Cnv3 (48)	DW-Cnv3 (48)
Layer 5	PW-Cnv1 (96)	PW-Cnv1 (96)
Layer 6	Max-Pool (2*2)	Max-Pool (2*2)
Layer 7	DW-Cnv3 (96)	DW-Cnv3 (96)
Layer 8	PW-Cnv1 (192)	PW-Cnv1 (192)
Layer 9	Max-Pool (2*2)	Max-Pool (2*2)
Layer 10	DW-Cnv3 (192)	DW-Cnv3 (192)
Layer 11	PW-Cnv1 (384)	PW-Cnv1 (384)
Layer 12	PW-Cnv1 (10)	PW-Cnv1 (10)
	Hardware Cost	
FFs(%)	27	38
BRAMs(%)	78	96
DSPs(%)	84	91
LUTs(%)	76	83
	Performances	
Frequency (MHz)	150	150
FPS	23	18
Latency (ms)	44	63.1
Power (W)	2.6	2.55
Energy Efficiency (J/image)	0.114	0.160

Table 1.
Results analysis.

the development of new IoT system designs that are decentralized from the cloud, where the majority of data processing operations are now handled. This tendency is much more visible in security-critical contexts, where IoT motes must make real-time judgments that cannot be transferred to cloud services because to the infrastructure network's interminable data transmission delays. Although microcontrollers provide the most programming freedom, their technology has reached its limits and cannot manage the increased computational power required by the upcoming generation of IoT devices. ASICs could satisfy this criterion, but they lack the programming/design flexibility that IoT systems demand. In this aspect, it is clear that reconfigurable platforms are an excellent implementation option for the upcoming generation of low-end IoT motes, as they provide unique competitive advantages such as flexibility through reconfigurable logic, versatility of hardware resources, high performance thanks to parallelism, and low power consumption with high security.

Acronyms and abbreviations

Nomenclature

FFs	Flip Flops
BRAMs	Block RAMs
DSPs	Digital Signal Processors
LUTs	Look Up Tables
FPS	Frame Per Seconds

Abbreviations

5/6G	Five and Six Generation Networks
IoT	Internet of Things
AI	Artificial Intelligence
FPGA	Field Programmable Gate Array
SoC	System on Chip
RFID	Radio Frequency Identification
IPV4	Internet Protocol version 4
LLN	Low-Power and Lossy Network
MAC	Medium Access Control
IDC	International Data Corporation
TRNG	True Random Number Generator
AES	Advanced Encryption Standard
SHA	Secure Hash Algorithm
ECC	Elliptical Curve Cryptography
MPU	Memory Protection Units
ASICs	Application-specific Integrated Circuits
DPR	Dynamic partial reconfiguration
SDR	Software-Defined Radio
FEC	Forward Error Correction
QoS	Quality of Service
QoR	Quality of Result

Author details

Seifeddine Messaoud[1*†], Rim Amdouni[1†], Adnen Albouchi[2†], Mohamed Ali Hajjaji[2†], Abdellatif Mtibaa[3†] and Mohamed Atri[4†]

1 Faculty of Sciences, Electronics and Microelectronics Lab., University of Monastir, Monastir, Tunisia

2 Electronics and Microelectronics Lab., ISAAT Sousse, University of Sousse, Sousse, Tunisia

3 Electronics and Microelectronics Lab., ENIM Monastir, University of Monastir, Monastir, Tunisia

4 College of Computer Science, King Khalid University, Abha, Saudi Arabia

*Address all correspondence to: seifeddine.messaoud@fsm.rnu.tn

† These authors contributed equally.

References

[1] Sinha BB, Dhanalakshmi R. Recent advancements and challenges of internet of things in smart agriculture: A survey. Future Generation Computer Systems. 2022;**126**:169-184

[2] Messaoud S, Bradai A, Hashim S, Bukhari R, Qung PTA, Ahmed OB, et al. A survey on machine learning in internet of things: Algorithms, strategies, and applications. Internet of Things. 2020;**12**: 100314 ISSN 2542-6605

[3] Di Martino B, Li KC, Yang LT, Esposito A. Trends and strategic researches in internet of everything. In: Internet of Everything. Singapore: Springer; 2018. pp. 1-12

[4] Posadas DV Jr. After the gold rush: The boom of the internet of things, and the busts of data-security and privacy. Fordham Intellectual Property, Media & Entertainment Law Journal. 2017; **28**:69

[5] Buyya R, Dastjerdi AV, editors. Internet of Things: Principles and Paradigms. Elsevier; 2016

[6] Rodríguez-Andina JJ, Valdes-Pena MD, Moure MJ. Advanced features and industrial applications of FPGAs—A review. IEEE Transactions on Industrial Informatics. 2015;**11**(4):853-864

[7] Messaoud S, Bouaafia S, Maraoui A, Ammari AC, Khriji L, Machhout M. Deep convolutional neural networks-based hardware–software on-chip system for computer vision application. Computers & Electrical Engineering. 2022;**98**:107671

[8] Sheng Z, Yang S, Yu Y, Vasilakos AV, McCann JA, Leung KK. A survey on the ietf protocol suite for the internet of things: Standards, challenges, and opportunities. IEEE Wireless Communications. 2013;**20**(6):91-98

[9] Shantharama P, Thyagaturu AS, Reisslein M. Hardware-accelerated platforms and infrastructures for network functions: A survey of enabling technologies and research studies. IEEE Access. 2020;**8**:132021-132085

[10] Lai L, Suda N, Chandra V. Cmsis-nn: Efficient neural network kernels for arm cortex-m cpus. arXiv preprint arXiv: 1801.06601. 19 Jan 2018;**1**. DOI: 10.48550/arXiv.1801.06601

[11] Alaba FA, Othman M, Hashem IAT, Alotaibi F. Internet of things security: A survey. Journal of Network and Computer Applications. 2017;**88**: 10-28

[12] Pinto S, Garlati C. "User mode interrupts—A must for securing embedded systems". In: Proceedings of the Embedded World Conference. Nuremberg, Bayern, Germany. 2019. pp. 505-510. Available: https://bringyourownit.com/2019/03/03/user-mode-interrupts-a-must-for-securing-embedded-systems/

[13] Shaikh FK, Zeadally S, Exposito E. Enabling technologies for green internet of things. IEEE Systems Journal. 2015; **11**(2):983-994

[14] Wang K, Wang Y, Sun Y, Guo S, Wu J. Green industrial internet of things architecture: An energy-efficient perspective. IEEE Communications Magazine. 2016;**54**(12):48-54

[15] Pena MDV, Rodriguez-Andina JJ, Manic M. The internet of things: The role of reconfigurable platforms. IEEE Industrial Electronics Magazine. 2017; **11**(3):6-19

[16] Tsai CW, Lai CF, Vasilakos AV. Future internet of things: Open issues and challenges. Wireless Networks. 2014;**20**(8):2201-2217

[17] Al-Kashoash HA, Kharrufa H, Al-Nidawi Y, Kemp AH. Congestion control in wireless sensor and 6LoWPAN networks: Toward the internet of things. Wireless Networks. 2019;**25**(8): 4493-4522

[18] Da Xu L, He W, Li S. Internet of things in industries: A survey. IEEE Transactions on Industrial Informatics. 2014;**10**(4):2233-2243

[19] Javed F, Afzal MK, Sharif M, Kim BS. Internet of things (IoT) operating systems support, networking technologies, applications, and challenges: A comparative review. IEEE Communications Surveys & Tutorials. 2018;**20**(3):2062-2100

[20] Lammie C, Olsen A, Carrick T, Azghadi MR. Low-power and high-speed deep FPGA inference Engines for Weed Classification at the edge. IEEE Access. 2019;7:51171-51184

[21] Molanes RF, Amarasinghe K, Rodriguez-Andina J, Manic M. Deep learning and reconfigurable platforms in the internet of things: Challenges and opportunities in algorithms and hardware. IEEE Industrial Electronics Magazine. 2018;**12**(2):36-49

[22] Luo T, Nagarajan SG. "Distributed anomaly detection using autoencoder neural networks in WSN for IoT". In: 2018 IEEE International Conference on Communications (ICC). Kansas City, MO, USA. 2018. pp. 1-6. doi: 10.1109/ICC.2018.8422402

[23] Koohang A, Sargent CS, Nord JH, Paliszkiewicz J. Internet of things (IoT): From awareness to continued use. International Journal of Information Management. 2022;**62**:102442

[24] Granjal J, Monteiro E, Silva JS. Security for the internet of things: A survey of existing protocols and open research issues. IEEE Communications Surveys & Tutorials. 2015;**17**(3): 1294-1312

[25] Chen K, Zhang S, Li Z, Zhang Y, Deng Q, Ray S, et al. Internet-of-things security and vulnerabilities: Taxonomy, challenges, and practice. Journal of Hardware and Systems Security. 2018; **2**(2):97-110

[26] Chen K, Zhang S, Li Z, Zhang Y, Deng Q, Ray S, et al. Internet-of-things security and vulnerabilities: Taxonomy, challenges, and practice. Journal of Hardware and Systems Security. 2018; **2**(2):97-110

[27] Pennekamp J, Henze M, Schmidt S, Niemietz P, Fey M, Trauth D, et al. Dataflow challenges in an internet of production: A security & privacy perspective. In: Proceedings of the ACM Workshop on Cyber-Physical Systems Security & Privacy. New York, NY, United States; 2019. pp. 27-38. DOI: 10.1145/3338499.3357357

[28] Benabdessalem R, Hamdi M, Kim TH. "A survey on security models, techniques, and tools for the internet of things". In: 2014 7th International Conference on Advanced Software Engineering and its Applications. Hainan, China. 2014. pp. 44-48. DOI: 10.1109/ASEA.2014.15

[29] Tan YS, Ko RKL, Holmes G. "Security and data accountability in distributed systems: A provenance survey". In: 2013 IEEE 10th International Conference on High Performance Computing and Communications & 2013 IEEE International Conference on

Embedded and Ubiquitous Computing. Zhangjiajie, China. 2013. pp. 1571-1578. DOI: 10.1109/HPCC.and.EUC.2013.221

[30] Restuccia F, D'Oro S, Melodia T. Securing the internet of things in the age of machine learning and software-defined networking. IEEE Internet of Things Journal. 2018;**5**(6):4829-4842

[31] Chatterjee S, Kar AK. "Regulation and Governance of the Internet of Things in India". Regulation and governance of the Internet of Things in India. 2018;**20**(5): pp. 399-412. DOI: 10.1108/DPRG-04-2018-0017

[32] Nisarga B, Peeters E. "System-Level Tamper Protection Using MSP MCUs." Dallas, Texas, United States: Texas Instruments; 2016. Available: https://e2e.ti.com/

[33] Kocher PC. Timing attacks on implementations of Diffie-Hellman, RSA, DSS, and other systems. In: Annual International Cryptology Conference. Berlin, Heidelberg: Springer; 1996. pp. 104-113

[34] Illuri B, Jose D, David S, Nagarjuan M. Machine learning based and reconfigurable architecture with a countermeasure for Side Channel attacks. In: Inventive Communication and Computational Technologies. Singapore: Springer; 2022. pp. 175-187

[35] Kamalinejad P, Mahapatra C, Sheng Z, Mirabbasi S, Leung VC, Guan YL. Wireless energy harvesting for the internet of things. IEEE Communications Magazine. 2015;**53**(6):102-108

[36] Alioto M, Shahghasemi M. The internet of things on its edge: Trends toward its tipping point. IEEE Consumer Electronics Magazine. 2017;**7**(1):77-87

[37] Rosello V, Portilla J, Riesgo T. "Ultra low power FPGA-based architecture for wake-up Radio in Wireless Sensor Networks." In: IECON 2011-37th Annual Conference of the IEEE Industrial Electronics Society. Melbourne, VIC, Australia. 2011. pp. 3826-3831. DOI: 10.1109/IECON.2011.6119933

[38] Monmasson E. Fpgas: Fundamentals, advanced features, and applications in industrial electronics [book news]. IEEE Industrial Electronics Magazine. 2017; **11**(2):73-74

[39] Gomes T, Salgado F, Pinto S, Cabral J, Tavares A. A 6LoWPAN accelerator for internet of things endpoint devices. IEEE Internet of Things Journal. 2017;**5**(1):371-377

[40] Rao M, Newe T, Grout I, Mathur A. An FPGA-based reconfigurable IPSec AH core with efficient implementation of SHA-3 for high speed IoT applications. Security and Communication Networks. 2016;**9**(16): 3282-3295

[41] Givehchi O, Landsdorf K, Simoens P, Colombo AW. Interoperability for industrial cyber-physical systems: An approach for legacy systems. IEEE Transactions on Industrial Informatics. 2017;**13**(6): 3370-3378

[42] Messaoud S, Bradai A, Ahmed OB, Quang PTA, Atri M, Hossain MS. Deep federated Q-learning-based network slicing for industrial IoT. IEEE Transactions on Industrial Informatics. 2020;**17**(8):5572-5582

[43] Qiu J, Wang J, Yao S, Guo K, Li B, Zhou E, et al. Going deeper with embedded fpga platform for convolutional neural network. In: ACM International Symposium on FPGA. New York, NY, United States. 2016. DOI: 10.1145/2847263.2847265 2016.

[44] Zhang J, Li J. Improving the performance of OpenCL-based FPGA accelerator for convolutional neural network. In: Proceedings of the 2017 ACM/SIGDA International Symposium on Field-Programmable Gate Arrays. New York, NY, United States: 2017. pp. 25-34. DOI: 10.1145/3020078.3021698

[45] Zhang X, et al. "Machine learning on FPGAs to face the IoT revolution." In: 2017 IEEE/ACM International Conference on Computer-Aided Design (ICCAD). Irvine, CA, USA. 2017. pp. 894-901 DOI: 10.1109/ICCAD.2017.8203875

[46] de la Piedra A, Braeken A, Touhafi A. "A performance comparison study of ECC and AES in commercial and research sensor nodes." In: Eurocon 2013. Zagreb, Croatia: IEEE. 2013. pp. 347-354. DOI: 10.1109/EUROCON.2013.6625007

[47] Xu X, Wang Y. "High speed true random number generator based on FPGA." In: 2016 International Conference on Information Systems Engineering (ICISE). Los Angeles, CA, USA. 2016. pp. 18-21. DOI: 10.1109/ICISE.2016.14

[48] Cicek I, Al Khas A. A new read–write collision-based SRAM PUF implemented on Xilinx FPGAs. Journal of Cryptographic Engineering. 2022; **2190-8516**:1-18. DOI: 10.1007/s13389-021-00281-8

[49] Schrijen GJ, Garlati C. Physical Unclonable Functions to the Rescue. In: Proceedings of the Embedded World. Nuremberg, Germany. 27 February–1 March 2018; 2018

[50] Johnson AP, Chakraborty RS, Mukhopadhyay D. A PUF-enabled secure architecture for FPGA-based IoT applications. IEEE Transactions on Multi-Scale Computing Systems. 2015; **1**(2):110-122

[51] Trimberger SMS. Three ages of FPGAs: A retrospective on the first thirty years of FPGA technology: This paper reflects on how Moore's law has driven the design of FPGAs through three epochs: The age of invention, the age of expansion, and the age of accumulation. IEEE Solid-State Circuits Magazine. 2018;**10**(2):16-29

[52] Ahmed I, Zhao S, Meijers J, Trescases O, Betz V. "Automatic BRAM Testing for Robust Dynamic Voltage Scaling for FPGAs." In: 2018 28th International Conference on Field Programmable Logic and Applications (FPL). Dublin, Ireland. 2018. pp. 68-687. DOI: 10.1109/FPL.2018.00020

[53] Karray F, Jmal MW, Garcia-Ortiz A, Abid M, Obeid AM. A comprehensive survey on wireless sensor node hardware platforms. Computer Networks. 2018; **144**:89-110

[54] Berder O, Sentieys O. Powwow: PowWow: Power Optimized Hardware/Software Framework for Wireless Motes. In: Proceedings of the 2010 23rd International Conference on Architecture of Computing Systems (ARCS). Hannover, Germany. 22–25 February 2010; pp. 1–5. Hannover, Germany: VDE; 2010. pp. 1-5

[55] Vera-Salas LA, Moreno-Tapia SV, Osornio-Rios RA, Romero-Troncoso Rd, "Reconfigurable Node Processing Unit for a Low-Power Wireless Sensor Network." In: 2010 International Conference on Reconfigurable Computing and FPGAs. Cancun, Mexico. 2010; pp. 173-178. DOI: 10.1109/ReConFig.2010.48

[56] Nyländen T, Boutellier J, Nikunen K, Hannuksela J, Silvén O. "Reconfigurable miniature sensor nodes for condition monitoring." In: 2012 International Conference on Embedded Computer Systems. Samos, Greece. 2012. pp. 113-119 DOI: 10.1109/SAMOS.2012.6404164

[57] MacGillivray C, Reinsel D. Worldwide Global DataSphere IoT Device and Data Forecast, 2019–2023 (IDC # US45066919). Framingham, MA, USA: International Data Corporation; 2019

[58] Huang L, Li DL, Wang KP, Gao T, Tavares A. A survey on performance optimization of high-level synthesis tools. Journal of Computer Science and Technology. 2020;**35**:697-720

[59] Coussy P, Gajski DD, Meredith M, Takach A. An introduction to high-level synthesis. IEEE Design & Test of Computers. 2009;**26**(4):8-17

[60] Zhang X, Hao C, Li Y, Chen Y, Xiong J, Hwu WM, et al. A bidirectional co-design approach to enable deep learning on IoT devices. Arxiv Preprint Arxiv:1905.08369. 2019;**1**

Chapter 3

Perspective Chapter: Internet of Things in Healthcare – New Trends, Challenges and Hurdles

Luis Muñoz-Saavedra, Francisco Luna-Perejón, Javier Civit-Masot and Elena Escobar-Linero

Abstract

Applied to health field, Internet of Things (IoT) systems provides continuous and ubiquitous monitoring and assistance, allowing the creation of valuable tools for diagnosis, health empowerment, and personalized treatment, among others. Advances in these systems follow different approaches, such as the integration of new protocols and standards, combination with artificial intelligence algorithms, application of big data processing methodologies, among others. These new systems and applications also should face different challenges when applying this kind of technology into health areas, such as the management of personal data sensed, integration with electronic health records, make sensing devices comfortable to wear, and achieve an accurate acquisition of the sensed data. The objective of this chapter is to present the state of the art, indicating the most current IoT trends applied to the health field, their contributions, technologies applied, and challenges faced.

Keywords: IoT systems, healthcare, eHealth, telehealth, medical support

1. Introduction

In recent years, the set of technologies encompassed under the name of the Internet of Things has experienced its greatest evolution and is currently approaching the slope of enlightenment of the hype cycle according to Gartner [1]. It has been applied in numerous areas, notably changing and improving the way in which different tasks and activities, both business and personal, are approached in daily life. Devices such as home assistants, home automation devices, and activity monitors are used more and more widely, providing information and functionalities that can be used quickly and easily.

One of the fields where there is more expectation about the application of this set of technologies is that related to healthcare and telehealth. Currently, there are several problems inherent in the health field that can be addressed thanks to the remote communication offered by the IoT. Advances in telehealth allow medical consultations and follow-up of patients in remote and isolated places, or with the limited mobility [2]. On the other hand, they enable the interconnection between health centers and remote systems that monitor elderly or disabled people who live alone or spend part of the time without company at all times, controlling vital signs or possible events such as falls that could endanger their lives [3].

Likewise, health services can be improved and optimized when health centers are provided with the capacity to integrate and interconnect devices that collect biomedical information with electronic health records [4]. Diagnosis, treatment, and follow-up in recovery from illnesses can be benefited in many cases by the continuous collection of these data [5], which complements the information obtained with specific observations that the medical professional can make during consultations, often limited in time. In addition, the data collected are a valuable source of information that can be used by Big Data and Artificial Intelligence applications to make new discoveries.

Although the advantages of these technologies applied to healthcare are clearly beneficial in many areas, there are also many aspects that make their implementation a challenging task. Due to the sensitive nature of the information, the technologies that must be implemented are those with characteristics that allow compliance with data privacy and security policies and standards [6]. On the other hand, they require health systems to have an appropriate infrastructure to accommodate these new technologies, as well as the adaptation of their protocols [7]. The training of health technicians, professionals, and patients to adapt them to these new systems is another relevant factor, and one that is related to usability and user experience [8].

Our purpose with this work is to analyze the evolution of IoT applied to healthcare and telehealth in recent years, the trends in application and what challenges currently exist. To address this objective, we will analyze the most relevant works in the recent years to draw conclusions about the global evolution of these technologies, check in more detail the problems they face, and identify whether there are standards, norms, or common complementary technologies to give a solution.

The rest of the article is divided as follows: In section two, the methodology of collection and analysis carried out are presented, detailing the aspects and characteristics on which we focus, in section three the results obtained are presented, and finally, the last section presents the conclusions.

2. Methodology

2.1 Search approach

The criteria established for the inclusion of studies in the analysis were that they had to be published in journals that appeared in the Journal Citation Reports (JCR) or in conferences included in prestigious journal editorials and digital libraries. The search engine used for the search engine was Google Scholar. The search was limited considering articles published in the last 5 completed years, that is, from January 2017 to December 2021 inclusive. The queries created to be used in the search engine consisted of the combinations of the term "IoT" with "healthcare," "ehealth," and "telehealth." The search for these terms was established to be carried out on the full text of each work, not only on the title and abstract. Since the purpose is to collect current advances and trends, we exclude from the analysis those articles with a scoping review nature. Similarly, we exclude studies that were written in a language other than English.

Regarding the results obtained after carrying out the search, the first three hundred were taken for each year in order of relevance, and one hundred for each of the three keywords considered in combination with "IoT," that is, "healthcare," "telehealth," and "ehealth." It was established to do a sequential filtering on the set, checking on each result that it meets each of the established inclusion requirements; otherwise, it will be discarded from the final set of results to be considered in the analysis. In the first place, the character of a high-impact journal paper or publication presented at a conference contemplated in prestigious digital libraries was verified, for which those results that were books or book chapters were also discarded. Next, the discarding of articles that were not written in English was applied. We considered that, to have a meaningful sample of current trends, a sample of 25 results would be taken for each combination of keyword and year. The most cited articles were used as a selection criterion. On this sample, studies consisting of bibliographic reviews were discarded. This last-filtering process was carried out firstly by analyzing the title of each publication and secondly by analyzing the abstract. Additionally, it was observed whether any study was replicated in the set or belonged to the same authors and the purpose was the same, in which case the study with the latest publication date was discarded.

For the initial purpose of filtering, basic information about these works was collected, specifically the title of the article, abstract, authors, access link to the publication, the number of citations, and character of the document according to Google Scholar. All the analyses of the information and filtering process were carried out jointly by the authors and verified between them.

2.2 Extracted information

Once the filtering process was carried out, the next step was to extract the relevant information for the analysis of the current situation, limitations, challenges, solutions, and current trends of IoT applied to healthcare and telemedicine. Mainly, we focus on extracting the most used communication technologies to provide solutions or address current problems in this area of research. Other relevant characteristics extracted were the scope of application from the point of view of what aspect of healthcare is intended to address or what characteristics of the infrastructure are intended to be addressed in the study. Related to the field, we also extract information on the technological aspects of the solution provided. Additionally, the country of the institution that supports each investigation was taken, in order to identify those countries that have the greatest impact worldwide in IoT for ehealth and telehealth fields.

3. Results and discussion

3.1 Results obtained after filtering process

After the filtering process, 186 [9–194] results were obtained, 69 with healthcare and another 69 with ehealth. The article analysis screening process resulted in a greater elimination of works related to telehealth, obtaining 48 results. This is mainly associated with the fact that after analyzing the title and abstract, it was detected that several articles did not fit in the field of healthcare using IoT. However, the cause of the greatest impact on the screening of results was the nature of the scoping review of several studies. Around eight publications per keyword and year on average were removed for this reason. Analyzing by year, the appearance of reviews was greater in the most recent years, discarding approximately 35 of 75 results of the year 2021. This reveals the progress of previous years and the current trend in analyzing the actual scope of IoT and limitations, which is consistent with the status of this area on hype-cycle curve.

3.2 Countries with contribution with higher impact

Figure 1 illustrates the percentage of publications according to the country of the institution of the corresponding author. In those cases in which the corresponding author was not reflected, the institution of the main author was taken as reference. The graph shows that the institutions with the greatest impact in recent years are India (19.3%) and China (13.4%). The rest of Asian countries (including those located in the Persian Gulf and Russia) contribute 25.8% to this statistic. Approximately 21% corresponds to institutions in European countries, approximately 6% to entities in Africa (mainly Egypt and Tunisia), and 4% to countries in Central and South America. The United States and Canada add 7.5% and Australia and New Zealand approximately 4.8%.

3.2.1 Analysis of scopes of higher impact

With the term "scope," we refer to the topics on which each study focuses on contributing to the area of IoT systems applied to ehealth and telemedicine. After the analysis, we have found studies whose scope is related to the provision of a health service, proposing models, system designs, and/or implementations of a complete system or a component of an IoT system. On the other hand, scopes focused on improving some characteristic of IoT systems that are relevant when applied to healthcare have also been identified.

Figure 2 illustrates the scopes in the analysis. It can be seen that the majority of studies focus on providing solutions for the field of monitoring. The reader should know that a division has been made in this scope, distinguishing between studies that explicitly indicated or from which the character of real-time monitoring could be clearly inferred. The total number of studies that fit this domain was 83. These data

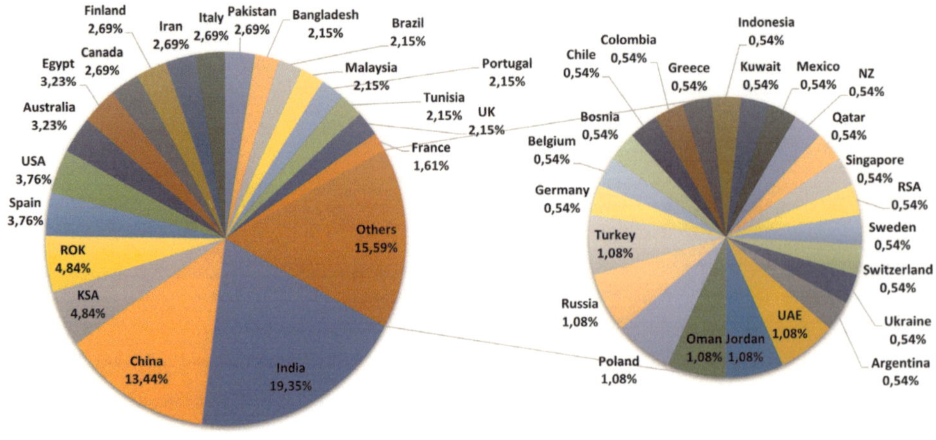

Figure 1.
Publications with more impact per country (N = 186).

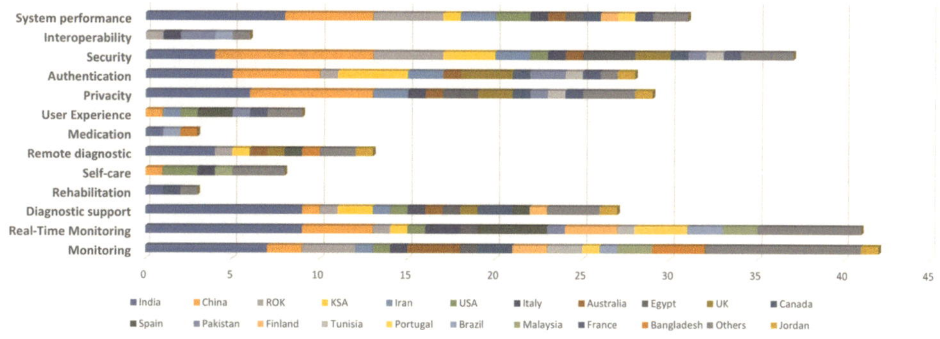

Figure 2.
Scope of the studies considered in the analysis.

show that the main purpose of the application of IoT to healthcare is to monitor patients, ubiquitously or integrated into rooms, for better control of vital signs or physiological parameters. This result is consistent with the main use for which IoT systems are used. Continuing with the analysis focused on areas of application, the next most common are those related to diagnosis. The creation of models that help the medical professional to give a diagnosis stands out mainly Machine Learning models that can be integrated into the system, sometimes complemented with architectures equipped with resources to apply Fog computing, as well as with the decentralization of processing with computing at the edge, trend that is currently increasing with the optimization of hardware and AI frameworks for model integration and consumption reduction [195]. Additionally, the applications are not restricted to the field of healthcare in the personal context, but also in the workplace [196]. To a lesser extent, we also find the use of IoT to facilitate the remote diagnosis of the patient. This last result may be related to the existing limitations to provide appropriate resources to remote centers or isolated areas that allow establishing reliable connections with sufficient transmission quality. The least relevant areas currently are self-care and remote rehabilitation. The first of these two areas mentioned was the one that had the greatest impact at the beginning of the use of IoT for healthcare, currently being on the slope of enlightenment or plateau of productivity in the hype-cycle curve. The low frequency of appearance of rehabilitation as a field of study may be due to the difficulty in carrying out rehabilitation tasks remotely.

Focusing on areas related to the improvement of system features, studies focused on maintaining the security of the IoT system are more frequent, that is, on avoiding transmission failures, network hacks, or data corruption. Additionally, we find studies focused on providing encryption protocols to ensure the authenticity and privacy of the patient. These last two scopes are often intrinsically related to system security. These results reveal the great challenges that exist in the integration of IoT systems in the Electronic Health Records of health systems: to be able to relate the data to the patient without compromising their privacy, as well as to manage the enormous amount of information collected avoiding losses, falsification information, and other security breaches. Very close in relevance we find the improvement of the performance of the system, that is, looking for better response times in the transmission and processing of information. Again, it is a challenging topic, especially in combination with equipping the system with authentication, privacy, and security protocols, which slows down IoT systems, which must be addressed for this type of system to be useful in terms of practicality.

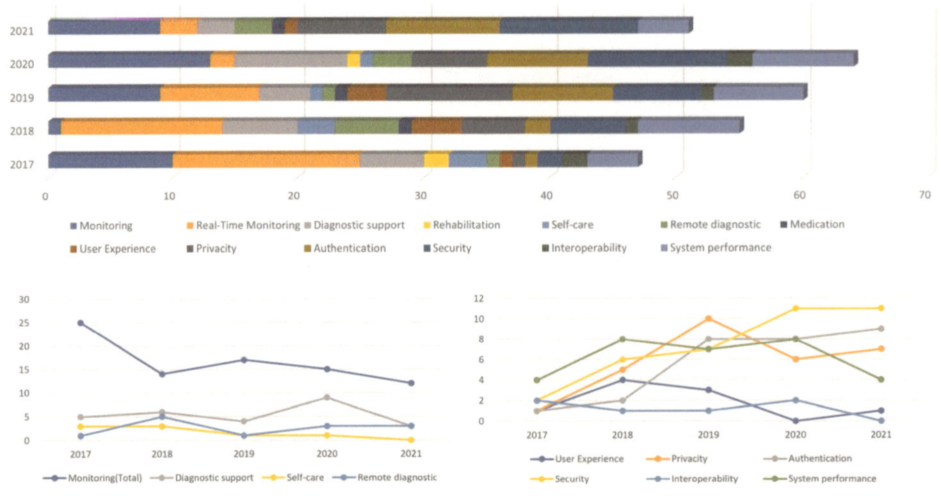

Figure 3.
Scope trends. At top, classification grouped by year analyzed in the study. At left, scopes related to the system's application approach. On the right, scopes focused on improving the characteristics of the IoT system.

In the results, we also observe that the studies related to the study of interoperability and the analysis of usability, user experience, and degree of acceptance have very little impact. These types of studies are not frequent, and yet, they address determining aspects in the adequacy of IoT systems to the health environment; its correct application depends on the fact that the implemented system is practical and perceived as useful. The little research on these issues may be the greatest limitation of these systems in the future.

Figure 3 shows the trend of these areas in the years considered for the analysis. Although a drop in the number of papers is perceived in 2021, this may be due to the fact that there has not yet been a stabilization in the number of citations on studies that contribute novelties to the field of research. Both graphs show how in recent years there has been a decrease in the impact of research aimed at providing solutions with IoT systems, and instead, there has been an increase in interest in research that focuses on improving some characteristics of the system architecture, mainly in security and privacy. This result again reveals indications of the situation of these systems in the hype-cycle curve, seeing reduced interest in deepen for new applications and consolidating their use by focusing on the greatest limitations that this set of technologies has, that is, aspects of security, privacy, and performance.

3.2.2 Interest in applied technologies

Paying attention to communication technologies, the charts in **Figure 4** reveal a varied set of alternatives. In these graphs, only those studies that have used and revealed the technologies applied in the implementation of IoT systems are taken into consideration. As a result, 87 articles were considered. The charts highlight the use of Bluetooth technology, in both its older versions and BLE, and Wi-Fi. Despite being a technology adapted to IoT, the use of LoRa is not frequent. GSM and GPRS technologies continue to be used, mainly because they have a greater network infrastructure for these technologies and because they are more in line with the user profile that

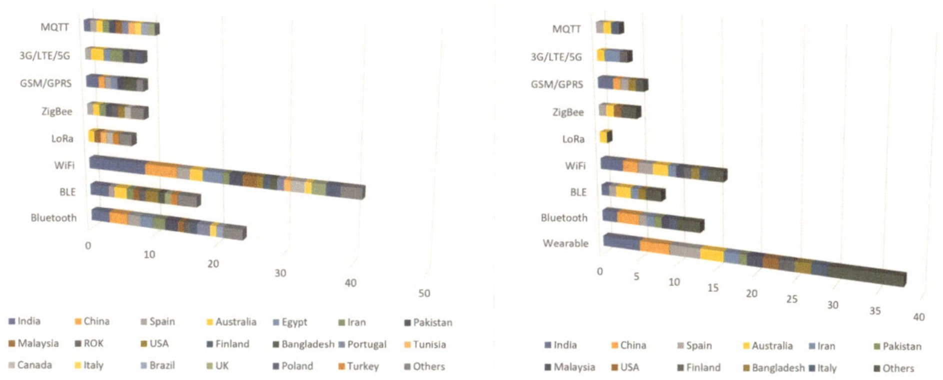

Figure 4.
Most common communication technologies in studies that carry out an implementation of an IoT system (N = 87). Top-right chart considers only those studies whose sensing acquisition devices are wearable (N = 36).

these systems are aimed at, mainly older people or those who are not familiar with new technologies.

Among the studies that reveal the communication technologies used, only 36 focus on the use of systems with exclusively wearable devices for data acquisition. In these studies, a lower use of MQTT is revealed in favor of the use of technologies such as Zigbee. Wi-Fi technology is still the most frequently used; however, there is a remarkable decrease.

The most frequent technologies identified in the analyzed studies include Machine Learning and Deep Learning for the fields of monitoring and diagnostic support. In the fields of security, privacy, and authentication, the use of Blockchain stands out. On the other hand, Fog computing and edge computing are the technologies for which the greatest interest is shown in the field of performance improvement. This is one of the most current trends, driven by systems equipped with more specialized hardware processing units [197].

4. Conclusions

The results obtained from the analysis of impact studies in recent years regarding IoT in healthcare show that Asian and Middle Eastern countries contribute to this area to a greater extent, especially India and China. With regard to the areas with the greatest impact today, we find the application to monitoring as the greatest representative. However, this type of study has reduced its relevance in recent years and instead has grown interest in the integration of security, privacy, and authentication measures to IoT systems, which gives indications of the stabilization of IoT technologies, and there is a tendency to investigate the improvement of the most important weaknesses. Infrequent and low-impact study topics are the analysis of the perceived usefulness of these systems, as well as interoperability, which may imply limitations and obstacles in the future in the implementation and use of these systems. The most used communication technologies are Bluetooth and Wi-Fi, with a smaller representation of technologies such as LoRa, Zigbee, and mobile data transfer technologies. The design and implementation of systems exclusively equipped with wearable acquisition devices is reduced. Machine Learning, Blockchain as well as edge and fog computing are the most trending technologies.

Acknowledgements

This work has been supported by "Fondo Europeo de Desarrollo Regional" (FEDER) and "Consejería de Economía, Conocimiento, Empresas y Universidad" of the Junta de Andalucía, under Programa Operativo FEDER 2014-2020 (Project US-1263715).

Conflict of interest

The authors declare no conflict of interest.

Author details

Luis Muñoz-Saavedra†, Francisco Luna-Perejón*†, Javier Civit-Masot†
and Elena Escobar-Linero†
University of Seville, Seville, Spain

*Address all correspondence to: fluna1@us.es

† These authors contributed equally.

References

[1] Lheureux B, et al. Hype Cycle for the Internet of Things [Internet], Gartner; 2021. Available from: https://www.gartner.com/en/documents/4005498 [Accessed 2022-02-15]

[2] Al-Majeed SS, Al-Mejibli IS, Karam J. Home telehealth by internet of things (IoT). In: 2015 IEEE 28th Canadian Conference on Electrical and Computer Engineering (CCECE). Piscataway, New Jersey, US: IEEE; 2015. pp. 609-613

[3] Luna-Perejón F, Muñoz-Saavedra L, Castellano-Domnguez JM, Domnguez-Morales M. IoT garment for remote elderly care network. Biomedical Signal Processing and Control. 2021;**69**: 102848

[4] Ganzha M, Paprzycki M, Pawłowski W, Szmeja P, Wasielewska K. Semantic interoperability in the internet of things: An overview from the INTER-IoT perspective. Journal of Network and Computer Applications. 2017;**81**:111-124

[5] Saheb T, Izadi L. Paradigm of IoT big data analytics in the healthcare industry: A review of scientific literature and mapping of research trends. Telematics and Informatics. 2019;**41**:70-85

[6] Gong T, Huang H, Li P, Zhang K, Jiang H. A medical healthcare system for privacy protection based on IoT. In: 2015 Seventh International Symposium on Parallel Architectures, Algorithms and Programming (PAAP). Piscataway, New Jersey, US: IEEE; 2015. p. 217-222.

[7] Tortorella GL, Fogliatto FS, Espôsto KF, Mac Cawley Vergara A, Vassolo R, Tlapa Mendoza D, et al. Measuring the effect of healthcare 4.0 implementation on hospitals' performance. Production Planning & Control. 2022;**33**(4):386-401

[8] El-Haddadeh R, Weerakkody V, Osmani M, Thakker D, Kapoor KK. Examining citizens' perceived value of internet of things technologies in facilitating public sector services engagement. Government Information Quarterly. 2019;**36**(2):310-320

[9] Satija U et al. Real-time signal quality-aware ECG telemetry system for IoT-based health care monitoring. IEEE Internet of Things Journal. 2017;**4**(3): 815-823

[10] Azimi I et al. HiCH: Hierarchical fog-assisted computing architecture for healthcare IoT. ACM Transactions on Embedded Computing Systems (TECS). 2017;**16**(5s):1-20

[11] Yang G et al. IoT-based remote pain monitoring system: From device to cloud platform. IEEE Journal of Biomedical and Health Informatics. 2017;**22**(6):1711-1719

[12] Lomotey RK et al. Wearable IoT data stream traceability in a distributed health information system. Pervasive and Mobile Computing. 2017;**40**:692-707

[13] Dubey H et al. Fog computing in medical internet-of-things: Architecture, implementation, and applications. In: Handbook of Large-Scale Distributed Computing in Smart Healthcare. Cham, Switzerland: Springer; 2017. pp. 281-321

[14] Hayati N, Suryanegara M. The IoT LoRa system design for tracking and monitoring patient with mental disorder. In: 2017 IEEE International Conference on Communication, Networks and Satellite (Comnetsat). Piscataway, New Jersey, US: IEEE; 2017. pp. 135-139

[15] Vora J et al. Home-based exercise system for patients using IoT enabled

smart speaker. In: IEEE 19th International Conference on e-Health Networking, Applications and Services (Healthcom). Piscataway, New Jersey, US: IEEE; 2017;**2017**:1-6

[16] Park K et al. An IoT system for remote monitoring of patients at home. Applied Sciences. 2017;7(3):260

[17] Ara A, Ara A. Case study: Integrating IoT, streaming analytics and machine learning to improve intelligent diabetes management system. In: International Conference on Energy, Communication, Data Analytics and Soft Computing (ICECDS). Piscataway, New Jersey, US: IEEE; 2017;**2017**:3179-3182

[18] Caporuscio M et al. Iot-enabled physical telerehabilitation platform. In: 2017 IEEE International Conference on Software Architecture Workshops (ICSAW). Piscataway, New Jersey, US: IEEE; 2017. pp. 112-119

[19] Antonio PO et al. Heat stroke detection system based in IoT. In: IEEE Second Ecuador Technical Chapters Meeting (ETCM). Piscataway, New Jersey, US: IEEE; 2017;**2017**:1-6

[20] Pham M et al. Delivering home healthcare through a cloud-based smart home environment (CoSHE). Future Generation Computer Systems. 2018;**81**: 129-140

[21] Liu C et al. Signal quality assessment and lightweight QRS detection for wearable ECG SmartVest system. IEEE Internet of Things Journal. 2018;**6**(2): 1363-1374

[22] Gurbeta L et al. A telehealth system for automated diagnosis of asthma and chronical obstructive pulmonary disease. Journal of the American Medical Informatics Association. 2018;**25**(9): 1213-1217

[23] Krishnan DSR et al. An IoT based patient health monitoring system. In: 2018 International Conference on Advances in Computing and Communication Engineering (ICACCE). Piscataway, New Jersey, US: IEEE; 2018. pp. 01-07

[24] Mulero R, Almeida A, Azkune G, Abril-Jiménez P, Waldmeyer MTA, Castrillo MP, et al. An IoT-aware approach for elderly-friendly cities. IEEE Access. 2018;**6**:7941-7957

[25] Stradolini F, et al. IoT for telemedicine practices enabled by an android™ application with cloud system integration. In: IEEE International Symposium on Circuits and Systems (ISCAS). Piscataway, New Jersey, US: IEEE; 2018;**2018**:1-5

[26] Ghosh D et al. Smart saline level monitoring system using ESP32 and MQTT-S. In: 2018 IEEE 20th International Conference on e-Health Networking, Applications and Services (Healthcom). Piscataway, New Jersey, US: IEEE; 2018. pp. 1-5

[27] Syed L et al. Telemammography: A novel approach for early detection of breast cancer through wavelets based image processing and machine learning techniques. In: Advances in Soft Computing and Machine Learning in Image Processing. Cham, Switzerland: Springer; 2018. pp. 149-183

[28] Stavrotheodoros S et al. A smart-home IoT infrastructure for the support of independent living of older adults. In: IFIP International Conference on Artificial Intelligence Applications and Innovations. Cham, Switzerland: Springer; 2018. pp. 238-249

[29] Manogaran G et al. Wearable IoT smart-log patch: An edge computing-based

Bayesian deep learning network system for multi access physical monitoring system. Sensors. 2019;**19**(13):3030

[30] Gia TN et al. Energy efficient fog-assisted IoT system for monitoring diabetic patients with cardiovascular disease. Future Generation Computer Systems. 2019;**93**:198-211

[31] Ozkan H et al. A portable wearable tele-ECG monitoring system. IEEE Transactions on Instrumentation and Measurement. 2019;**69**(1):173-182

[32] Saadeh W et al. A patient-specific single sensor IoT-based wearable fall prediction and detection system. IEEE Transactions on Neural Systems and Rehabilitation Engineering. 2019;**27**(5):995-1003

[33] Gutiérrez-Madroñal L, La Blunda L, Wagner MF, Medina-Bulo I. Test event generation for a fall-detection IoT system. IEEE Internet of Things Journal. 2019;**6**(4):6642-6651

[34] Yee LM et al. Internet of things (IoT) fall detection using wearable sensor. Journal of Physics: Conference Series. 2019;**1372**:012048

[35] Islam MR et al. Design and implementation of low cost smart syringe pump for telemedicine and healthcare. In: 2019 International Conference on Robotics, Electrical and Signal Processing Techniques (ICREST). Piscataway, New Jersey, US: IEEE; 2019. pp. 440-444

[36] Al-Kababji A et al. IoT-based fall and ECG monitoring system: Wireless communication system based firebase realtime database. In: IEEE SmartWorld, Ubiquitous Intelligence & Computing, Advanced & Trusted Computing, Scalable Computing & Communications, Cloud & big Data Computing, Internet of People and Smart City Innovation (Smart-World/SCALCOM/UIC/ATC/CBDCom/IOP/SCI). Piscataway, New Jersey, US: IEEE; 2019;**2019**:1480-1485

[37] Joseph S et al. IOT based remote heartbeat monitoring. In: 2019 International Conference on Advances in Computing, Communication and Control (ICAC3). Piscataway, New Jersey, US: IEEE; 2019. pp. 1-5

[38] Zhao P et al. Towards deep learning-based detection scheme with raw ECG signal for wearable telehealth systems. In: 2019 28th International Conference on Computer Communication and Networks (ICCCN). Piscataway, New Jersey, US: IEEE; 2019. pp. 1-9

[39] Shaik MS et al. Detection of FITS seizure by Alexa using IoT. In: 2019 International Conference on Vision Towards Emerging Trends in Communication and Networking (ViTECoN). Piscataway, New Jersey, US: IEEE; 2019. pp. 1-4

[40] Singh RP et al. Internet of things (IoT) applications to fight against COVID-19 pandemic. Diabetes & Metabolic Syndrome: Clinical Research & Reviews. 2020;**14**(4):521-524

[41] Rahman MS, Peeri NC, Shrestha N, Zaki R, Haque U, Ab Hamid SH. Defending against the novel coronavirus (COVID-19) outbreak: How can the internet of things (IoT) help to save the world? Health Policy and Technology. 2020;**9**(2):136

[42] Islam M et al. Development of smart healthcare monitoring system in IoT environment. SN Computer Science. 2020;**1**(3):1-11

[43] Otoom M et al. An IoT-based framework for early identification and

monitoring of COVID-19 cases. Biomedical Signal Processing and Control. 2020;**62**:102149

[44] Siriwardhana Y, De Alwis C, et al. The fight against the COVID-19 pandemic with 5G technologies. IEEE Engineering Management Review. 2020; **48**(3):72-84

[45] Hoffman DA. Increasing access to care: Telehealth during COVID-19. Journal of Law and the Biosciences. 2020;**7**(1):lsaa043

[46] Farahani B et al. Towards collaborative intelligent IoT eHealth: From device to fog, and cloud. Microprocessors and Microsystems. 2020;**72**:102938

[47] Abdelmoneem RM et al. Mobility-aware task scheduling in cloud-fog IoT-based healthcare architectures. Computer Networks. 2020;**179**:107348

[48] Muneer A et al. Smart health monitoring system using IoT based smart fitness mirror. Telkomnika. 2020; **18**(1):317-331

[49] Asadzadeh A et al. Information technology in emergency management of COVID-19 outbreak. Informatics in Medicine Unlocked. 2020;**21**:100475

[50] Gunasekeran DV et al. Digital health during COVID-19: Lessons from operationalising new models of care in ophthalmology. The Lancet Digital Health. 2021;**3**(2):e124-e134

[51] Ahmad RW et al. The role of blockchain technology in telehealth and telemedicine. International Journal of Medical Informatics. 2021; **148**:104399

[52] Arfi WB et al. Understanding acceptance of eHealthcare by IoT natives and IoT immigrants: An integrated model of UTAUT, perceived risk, and financial cost. Technological Forecasting and Social Change. 2021;**163**: 120437

[53] Honar Pajooh H et al. Hyperledger fabric blockchain for securing the edge internet of things. Sensors. 2021; **21**(2):359

[54] Ullah SMA et al. Scalable telehealth services to combat novel coronavirus (COVID-19) pandemic. SN Computer Science. 2021;**2**(1):1-8

[55] Wang W et al. Blockchain-assisted handover authentication for intelligent telehealth in multi-server edge computing environment. Journal of Systems Architecture. 2021;**115**: 102024

[56] Mukati N et al. Healthcare assistance to COVID-19 patient using internet of things (IoT) enabled technologies. Materials Today: Proceedings. 2021. Available from: https://doi.org/10.1016/j.matpr.2021.07.379

[57] Wu T et al. An autonomous wireless body area network implementation towards IoT connected healthcare applications. IEEE Access. 2017;**5**: 11413-11422

[58] Li C et al. The IoT-based heart disease monitoring system for pervasive healthcare service. Procedia Computer Science. 2017;**112**:2328-2334

[59] Mora H, Gil D, Terol RM, Azorín J, Szymanski J. An IoT-based computational framework for healthcare monitoring in mobile environments. Sensors. 2017;**17**(10):2302

[60] Jabbar S et al. Semantic interoperability in heterogeneous IoT

infrastructure for healthcare. Wireless Communications and Mobile Computing. 2017;**2017**:10

[61] Ullah F et al. Semantic interoperability for big-data in heterogeneous IoT infrastructure for healthcare. Sustainable Cities and Society. 2017;**34**:90-96

[62] Khan SF. Health care monitoring system in internet of things (IoT) by using RFID. In: 2017 6th International Conference on Industrial Technology and Management (ICITM). IEEE; 2017. pp. 198-204

[63] Sood SK, Mahajan I. Wearable IoT sensor based healthcare system for identifying and controlling chikungunya virus. Computers in Industry. 2017;**91**: 33-44

[64] Dziak D et al. IoT-based information system for healthcare application: Design methodology approach. Applied Sciences. 2017;**7**(6):596

[65] Laplante PA et al. Building caring healthcare systems in the internet of things. IEEE Systems Journal. 2017; **12**(3):3030-3037

[66] Mezghani E et al. A model-driven methodology for the design of autonomic and cognitive IoT-based systems: Application to healthcare. IEEE Transactions on Emerging Topics in Computational Intelligence. 2017;**1**(3): 224-234

[67] Bhatia M, Sood SK. A comprehensive health assessment framework to facilitate IoT-assisted smart workouts: A predictive healthcare perspective. Computers in Industry. 2017;**92**:50-66

[68] Strielkina A et al. Modelling of healthcare IoT using the queueing theory. In: 9th IEEE International Conference on Intelligent Data Acquisition and Advanced Computing Systems: Technology and Applications (IDAACS). Piscataway, New Jersey, US: IEEE; 2017;**2**:849-852

[69] He S et al. Proactive personalized services through fog-cloud computing in large-scale IoT-based healthcare application. China Communications. 2017;**14**(11):1-16

[70] Jaiswal K et al. IoT-cloud based framework for patient's data collection in smart healthcare system using raspberry-pi. In: International Conference on Electrical and Computing Technologies and Applications (ICECTA). Piscataway, New Jersey, US: IEEE; 2017;**2017**:1-4

[71] Elhoseny M, Ramrez-González G, Abu-Elnasr OM, Shawkat SA, Arunkumar N, Farouk A. Secure medical data transmission model for IoT-based healthcare systems. IEEE Access. 2018;**6**: 20596-20608

[72] Verma P, Sood SK. Fog assisted-IoT enabled patient health monitoring in smart homes. IEEE Internet of Things Journal. 2018;**5**(3):1789-1796

[73] Pace P et al. An edge-based architecture to support efficient applications for healthcare industry 4.0. IEEE Transactions on Industrial Informatics. 2018;**15**(1):481-489

[74] Kumar PM et al. Cloud and IoT based disease prediction and diagnosis system for healthcare using fuzzy neural classifier. Future Generation Computer Systems. 2018;**86**:527-534

[75] Luo E et al. Privacyprotector: Privacy-protected patient data collection in IoT-based healthcare systems. IEEE Communications Magazine. 2018;**56**(2): 163-168

[76] Mahmud R et al. Cloud-fog interoperability in IoT-enabled healthcare solutions. In: Proceedings of the 19th International Conference on Distributed Computing and Networking. New York, NY, USA: Association for Computing Machinery; 2018. pp. 1-10

[77] Verma P, Sood SK. Cloud-centric IoT based disease diagnosis healthcare framework. Journal of Parallel and Distributed Computing. 2018;**116**:27-38

[78] Woo MW et al. A reliable IoT system for personal healthcare devices. Future Generation Computer Systems. 2018;**78**: 626-640

[79] Min M et al. Learning-based privacy-aware offloading for healthcare IoT with energy harvesting. IEEE Internet of Things Journal. 2018;**6**(3): 4307-4316

[80] Tao H et al. Secured data collection with hardware-based ciphers for IoT-based healthcare. IEEE Internet of Things Journal. 2018;**6**(1):410-420

[81] Alhussein M et al. Cognitive IoT-cloud integration for smart healthcare: Case study for epileptic seizure detection and monitoring. Mobile Networks and Applications. 2018;**23**(6):1624-1635

[82] Catherwood PA et al. A community-based IoT personalized wireless healthcare solution trial. IEEE Journal of Translational Engineering in Health and Medicine. 2018;**6**:1-13

[83] Subasi A et al. IoT based mobile healthcare system for human activity recognition. In: 2018 15th Learning and Technology Conference (L&T). Piscataway, New Jersey, USA: IEEE; 2018;**2018**:29-34

[84] Srinivasa K et al. Data analytics assisted internet of things towards building intelligent healthcare monitoring systems: Iot for healthcare. Journal of Organizational and End User Computing (JOEUC). 2018;**30**(4): 83-103

[85] Badr S et al. Multi-tier blockchain framework for IoT-EHRs systems. Procedia Computer Science. 2018;**141**: 159-166

[86] Martínez-Caro E et al. Healthcare service evolution towards the internet of things: An end-user perspective. Technological Forecasting and Social Change. 2018;**136**:268-276

[87] Wang K et al. Adaptive and fault-tolerant data processing in healthcare IoT based on fog computing. IEEE Transactions on Network Science and Engineering. 2018;**7**(1):263-273

[88] Dwivedi AD et al. A decentralized privacy-preserving healthcare blockchain for IoT. Sensors. 2019;**19**(2):326

[89] Yang Y et al. Privacy-preserving smart IoT-based healthcare big data storage and self-adaptive access control system. Information Sciences. 2019;**479**: 567-592

[90] Kaur P et al. A healthcare monitoring system using random forest and internet of things (IoT). Multimedia Tools and Applications. 2019;**78**(14): 19905-19916

[91] Deebak BD, Al-Turjman F, et al. An authentic-based privacy preservation protocol for smart e-healthcare systems in IoT. IEEE Access. 2019;**7**:135632-135649

[92] Azimi I et al. Missing data resilient decision-making for healthcare IoT through personalization: A case study on maternal health. Future Generation Computer Systems. 2019;**96**:297-308

[93] Alraja MN et al. The effect of security, privacy, familiarity, and trust on users' attitudes toward the use of the IoT-based healthcare: The mediation role of risk perception. IEEE Access. 2019;7: 111341-111354

[94] Dautov R et al. Hierarchical data fusion for smart healthcare. Journal of Big Data. 2019;**6**(1):1-23

[95] Shahidul Islam M et al. Monitoring of the human body signal through the internet of things (IoT) based LoRa wireless network system. Applied Sciences. 2019;**9**(9):1884

[96] Sharma G, Kalra S. A lightweight user authentication scheme for cloud-IoT based healthcare services. Iranian Journal of Science and Technology, Transactions of Electrical Engineering. 2019;**43**(1):619-636

[97] Elmisery AM et al. A new computing environment for collective privacy protection from constrained healthcare devices to IoT cloud services. Cluster Computing. 2019;**22**(1):1611-1638

[98] Srivastava G et al. A light and secure healthcare blockchain for iot medical devices. In: IEEE Canadian Conference of Electrical and Computer Engineering (CCECE). Piscataway, New Jersey, US: IEEE; 2019;**2019**:1-5

[99] Tang W et al. Secure data aggregation of lightweight E-healthcare IoT devices with fair incentives. IEEE Internet of Things Journal. 2019;**6**(5): 8714-8726

[100] Abou-Nassar EM et al. DITrust chain: Towards blockchain-based trust models for sustainable healthcare IoT systems. IEEE Access. 2020;**8**: 111223-111238

[101] Rathee G et al. A hybrid framework for multimedia data processing in IoT-healthcare using blockchain technology. Multimedia Tools and Applications. 2020;**79**(15):9711-9733

[102] Muthu B, Sivaparthipan C, Manogaran G, Sundarasekar R, Kadry S, Shanthini A, et al. IOT based wearable sensor for diseases prediction and symptom analysis in healthcare sector. Peer-to-peer Networking and Applications. 2020;**13**(6):2123-2134

[103] Haghi M et al. A flexible and pervasive IoT-based healthcare platform for physiological and environmental parameters monitoring. IEEE Internet of Things Journal. 2020;7(6):5628-5647

[104] Celesti A et al. Blockchain-based healthcare workflow for tele-medical laboratory in federated hospital IoT clouds. Sensors. 2020;**20**(9):2590

[105] Bharathi R et al. Energy efficient clustering with disease diagnosis model for IoT based sustainable healthcare systems. Sustainable Computing: Informatics and Systems. 2020;**28**: 100453

[106] Wu T et al. A rigid-flex wearable health monitoring sensor patch for IoT-connected healthcare applications. IEEE Internet of Things Journal. 2020;7(8): 6932-6945

[107] Li J et al. A secured framework for sdn-based edge computing in IOT-enabled healthcare system. IEEE Access. 2020;**8**:135479-135490

[108] Ullah A et al. Fog-assisted secure healthcare data aggregation scheme in IoT-enabled WSN. Peer-to-Peer Networking and Applications. 2020; **13**(1):163-174

[109] Gope P et al. A secure IoT-based modern healthcare system with fault-

tolerant decision making process. IEEE Journal of Biomedical and Health Informatics. 2020;**25**(3):862-873

[110] Aujla GS, Jindal A. A decoupled blockchain approach for edge-envisioned IoT-based healthcare monitoring. IEEE Journal on Selected Areas in Communications. 2020;**39**(2): 491-499

[111] Wang X, Cai S. Secure healthcare monitoring framework integrating NDN-based IoT with edge cloud. Future Generation Computer Systems. 2020; **112**:320-329

[112] Guo X et al. A new data clustering strategy for enhancing mutual privacy in healthcare IoT systems. Future Generation Computer Systems. 2020; **113**:407-417

[113] Satpathy S et al. A new healthcare diagnosis system using an IoT-based fuzzy classifier with FPGA. The Journal of Supercomputing. 2020;**76**(8):5849-5861

[114] Sun Y et al. PMRSS: Privacy-preserving medical record searching scheme for intelligent diagnosis in IoT healthcare. IEEE Transactions on Industrial Informatics. 2021;**18**(3): 1981-1990

[115] Onasanya A, Elshakankiri M. Smart integrated IoT healthcare system for cancer care. Wireless Networks. 2021; **27**(6):4297-4312

[116] El Zouka HA, Hosni MM. Secure IoT communications for smart healthcare monitoring system. Internet of Things. 2021;**13**:100036

[117] Elayan H et al. Digital twin for intelligent context-aware iot healthcare systems. IEEE Internet of Things Journal. 2021;**8**(23): 16749-16757

[118] Alzubi JA. Blockchain-based Lamport Merkle digital signature: Authentication tool in IoT healthcare. Computer Communications. 2021;**170**: 200-208

[119] Wu TY et al. Improved authenticated key agreement scheme for fog-driven IoT healthcare system. Security and communication. Wireless and communication Networks. 2021;**2021**:19

[120] Mukherjee R et al. IoT-cloud based healthcare model for COVID-19 detection: An enhanced k-nearest neighbour classifier based approach. Computing. 2021;**1-21**

[121] Rajavel R et al. IoT-based smart healthcare video surveillance system using edge computing. Journal of Ambient Intelligence and Humanized Computing. 2021;**13**:3195-3207

[122] Wu F et al. Edge-based hybrid system implementation for long-range safety and healthcare IoT applications. IEEE Internet of Things Journal. 2021; **8**(12):9970-9980

[123] Poongodi M et al. Smart healthcare in smart cities: Wireless patient monitoring system using IoT. The Journal of Supercomputing. 2021;**77**(11): 12230-12255

[124] Magsi H et al. A novel adaptive battery-aware algorithm for data transmission in IoT-based healthcare applications. Electronics. 2021;**10**(4):367

[125] de Morais Barroca Filho I et al. An IoT-based healthcare platform for patients in ICU beds during the COVID-19 outbreak. IEEE Access. 2021;**9**: 27262-27277

[126] Rifi N et al. Towards using blockchain technology for eHealth data

access management. In: 2017 Fourth International Conference on Advances in Biomedical Engineering (ICABME). Piscataway, New Jersey, US: IEEE; 2017. pp. 1-4

[127] Al-Hamadi H, Chen R. Trust-based decision making for health IoT systems. IEEE Internet of Things Journal. 2017; **4**(5):1408-1419

[128] Gia TN et al. Low-cost fog-assisted health-care IoT system with energy-efficient sensor nodes. In: 2017 13th International Wireless Communications and Mobile Computing Conference (IWCMC). Piscataway, New Jersey, US: IEEE; 2017. pp. 1765-1770

[129] Mdhaffar A et al. IoT-based health monitoring via LoRaWAN. In: IEEE EUROCON 2017-17th International Conference on Smart Technologies. Piscataway, New Jersey, US: IEEE; 2017. pp. 519-524

[130] Gupta PK et al. A novel and secure IoT based cloud centric architecture to perform predictive analysis of users activities in sustainable health centres. Multimedia Tools and Applications. 2017;**76**(18):18489-18512

[131] Neyja M et al. An IoT-based e-health monitoring system using ECG signal. In: GLOBECOM 2017–2017 IEEE Global Communications Conference. Piscataway, New Jersey, US: IEEE; 2017. pp. 1-6

[132] Domingues MF et al. Insole optical fiber sensor architecture for remote gait analysis—An e-health solution. IEEE Internet of Things Journal. 2017;**6**(1): 207-214

[133] Rathore MM et al. Hadoop-based intelligent care system (HICS) analytical approach for big data in IoT. ACM Transactions on Internet Technology (TOIT). 2017;**18**(1):1-24

[134] Raj C et al. HEMAN: Health monitoring and nous: An IoT based e-health care system for remote telemedicine. In: 2017 International Conference on Wireless Communications, Signal Processing and Networking (WiSPNET). Piscataway, New Jersey, US: IEEE; 2017. pp. 2115-2119

[135] Ali S, Ghazal M. Real-time heart attack mobile detection service (RHAMDS): An IoT use case for software defined networks. In: 2017 IEEE 30th Canadian Conference on Electrical and Computer Engineering (CCECE). Piscataway, New Jersey, US: IEEE; 2017. pp. 1-6

[136] Buyukakkaslar MT et al. LoRaWAN as an e-health communication technology. In: 2017 IEEE 41st Annual Computer Software and Applications Conference (COMPSAC). Piscataway, New Jersey, US: IEEE; 2017;**2**:310-313

[137] Chatterjee P et al. IoT-based decision support system for intelligent healthcare—Applied to cardiovascular diseases. In: 2017 7th International Conference on Communication Systems and Network Technologies (CSNT). Piscataway, New Jersey, US: IEEE; 2017. pp. 362-366

[138] Cabra J et al. An IoT approach for wireless sensor networks applied to e-health environmental monitoring. In: 2017 IEEE International Conference on Internet of Things (iThings) and IEEE Green Computing and Communications (GreenCom) and IEEE Cyber, Physical and Social Computing (CPSCom) and IEEE Smart Data (SmartData). Piscataway, New Jersey, US: IEEE; 2017. pp. 578-583

[139] Budida DAM, Mangrulkar RS. Design and implementation of smart HealthCare system using IoT. In: 2017 International Conference on Innovations in Information, Embedded and Communication Systems (ICIIECS). Piscataway, New Jersey, US: IEEE; 2017. pp. 1-7

[140] Rahmani AM et al. Exploiting smart e-health gateways at the edge of healthcare internet-of-things: A fog computing approach. Future Generation Computer Systems. 2018;**78**: 641-658

[141] Farahani B et al. Towards fog-driven IoT eHealth: Promises and challenges of IoT in medicine and healthcare. Future Generation Computer Systems. 2018;**78**:659-676

[142] Kumari A et al. Fog computing for healthcare 4.0 environment: Opportunities and challenges. Computers & Electrical Engineering. 2018;**72**:1-13

[143] Rodrigues JJ, Segundo DBDR, Junqueira HA, Sabino MH, Prince RM, Al-Muhtadi J, et al. Enabling technologies for the internet of health things. IEEE Access. 2018;**6**: 13129-13141

[144] Zhang X, Poslad S. Blockchain support for flexible queries with granular access control to electronic medical records (EMR). In: 2018 IEEE International Conference on Communications (ICC). Piscataway, New Jersey, US: IEEE; 2018. pp. 1-6

[145] Almulhim M, Zaman N. Proposing secure and lightweight authentication scheme for IoT based E-health applications. In: 2018 20th International Conference on Advanced Communication Technology (ICACT). Piscataway, New Jersey, US: IEEE; 2018. pp. 481-487

[146] Bayo-Monton JL et al. Wearable sensors integrated with internet of things for advancing eHealth care. Sensors. 2018;**18**(6):1851

[147] Chen X et al. Dynamic power management and adaptive packet size selection for IoT in e-healthcare. Computers & Electrical Engineering. 2018;**65**:357-375

[148] Santos GL et al. Analyzing the availability and performance of an e-health system integrated with edge, fog and cloud infrastructures. Journal of Cloud Computing. 2018;**7**(1):1-22

[149] Naranjo-Hernández D et al. Smart vest for respiratory rate monitoring of COPD patients based on non-contact capacitive sensing. Sensors. 2018;**18**(7): 2144

[150] Monteiro K et al. Developing an e-health system based on IoT, fog and cloud computing. In: 2018 IEEE/ACM International Conference on Utility and Cloud Computing Companion (UCC Companion). Piscataway, New Jersey, US: IEEE; 2018. pp. 17-18

[151] Pasha M, Shah SMW. Framework for E-health systems in IoT-based environments. Wireless Communications and Mobile Computing. 2018;**2018**:11

[152] Santamaria AF et al. A real IoT device deployment for e-health applications under lightweight communication protocols, activity classifier and edge data filtering. Computer Communications. 2018;**128**: 60-73

[153] Abdel-Basset M et al. A novel intelligent medical decision support

model based on soft computing and IoT. IEEE Internet of Things Journal. 2019; **7**(5):4160-4170

[154] Jia X et al. Authenticated key agreement scheme for fog-driven IoT healthcare system. Wireless Networks. 2019;**25**(8):4737-4750

[155] Aghili SF et al. LACO: Lightweight three-factor authentication, access control and ownership transfer scheme for e-health systems in IoT. Future Generation Computer Systems. 2019;**96**: 410-424

[156] Vilela PH et al. Performance evaluation of a fog-assisted IoT solution for e-health applications. Future Generation Computer Systems. 2019;**97**: 379-386

[157] Saha R et al. Privacy ensured e-healthcare for fog-enhanced IoT based applications. IEEE Access. 2019;**7**: 44536-44543

[158] Rath M, Pattanayak B. Technological improvement in modern health care applications using internet of things (IoT) and proposal of novel health care approach. International Journal of Human Rights in Healthcare. 2018;**12**: 148-162

[159] Debauche O et al. Fog IoT for health: A new architecture for patients and elderly monitoring. Procedia Computer Science. 2019;**160**:289-297

[160] Hasan M et al. Real-time healthcare data transmission for remote patient monitoring in patch-based hybrid OCC/ BLE networks. Sensors. 2019;**19**(5):1208

[161] Kaw JA et al. A reversible and secure patient information hiding system for IoT driven e-health. International Journal of Information Management. 2019;**45**:262-275

[162] Almulhim M et al. A lightweight and secure authentication scheme for IoT based e-health applications. International Journal of Computer Science and Network Security. 2019; **19**(1):107-120

[163] Hossein KM et al. Blockchain-based privacy-preserving healthcare architecture. In: 2019 IEEE Canadian Conference of Electrical and Computer Engineering (CCECE). Piscataway, New Jersey, US: IEEE; 2019. pp. 1-4

[164] Ben Hassen H et al. An E-health system for monitoring elderly health based on internet of things and fog computing. Health Information Science and Systems. 2019;**7**(1):1-9

[165] Alassaf N, Gutub A. Simulating light-weight-cryptography implementation for IoT healthcare data security applications. International Journal of E-Health and Medical Communications (IJEHMC). 2019; **10**(4):1-15

[166] Celesti A et al. How to develop IoT cloud e-health systems based on FIWARE: A lesson learnt. Journal of Sensor and Actuator Networks. 2019; **8**(1):7

[167] Ali R et al. Q-learning-enabled channel access in next-generation dense wireless networks for IoT-based eHealth systems. EURASIP Journal on Wireless Communications and Networking. 2019; **2019**(1):1-12

[168] Tuli S et al. HealthFog: An ensemble deep learning based smart healthcare system for automatic diagnosis of heart diseases in integrated IoT and fog computing environments. Future Generation Computer Systems. 2020; **104**:187-200

[169] Hamza R et al. A privacy-preserving cryptosystem for IoT E-healthcare. Information Sciences. 2020;**527**:493-510

[170] Jamil F et al. Towards a remote monitoring of patient vital signs based on IoT-based blockchain integrity management platforms in smart hospitals. Sensors. 2020;**20**(8):2195

[171] Ray PP et al. Blockchain for IoT-based healthcare: Background, consensus, platforms, and use cases. IEEE Systems Journal. 2020;**15**(1): 85-94

[172] Vedaei SS et al. COVID-SAFE: An IoT-based system for automated health monitoring and surveillance in post-pandemic life. IEEE Access. 2020;**8**:188538

[173] Khan MA et al. A secure framework for authentication and encryption using improved ECC for IoT-based medical sensor data. IEEE Access. 2020;**8**: 52018-52027

[174] Patan R et al. Smart healthcare and quality of service in IoT using grey filter convolutional based cyber physical system. Sustainable Cities and Society. 2020;**59**:102141

[175] Cai G et al. Design of an MISO-SWIPT-aided code-index modulated multi-carrier M-DCSK system for e-health IoT. IEEE Journal on Selected Areas in Communications. 2020;**39**(2):311-324

[176] Yew HT et al. Iot based real-time remote patient monitoring system. In: 16th IEEE International Colloquium on Signal Processing & its Applications (CSPA). Piscataway, New Jersey, US: IEEE; 2020;**2020**:176-179

[177] Almogren A et al. Ftm-iomt: Fuzzy-based trust management for preventing sybil attacks in internet of medical things. IEEE Internet of Things Journal. 2020;**8**(6):4485-4497

[178] Uddin MA et al. Blockchain leveraged decentralized IoT eHealth framework. Internet of Things. 2020;**9**: 100159

[179] Chowdhury MZ et al. A new 5g ehealth architecture based on optical camera communication: An overview, prospects, and applications. IEEE Consumer Electronics Magazine. 2020; **9**(6):23-33

[180] Rub JNS, Gondim PRL. Interoperable internet of medical things platform for e-health applications. International Journal of Distributed Sensor Networks. 2020;**16**(1): 1550147719889591

[181] Zhang L et al. A multi-stage stochastic programming-based offloading policy for fog enabled IoT-eHealth. IEEE Journal on Selected Areas in Communications. 2020;**39**(2):411-425

[182] Ghosh A et al. Energy-efficient IoT-health monitoring system using approximate computing. Internet of Things. 2020;**9**:100166

[183] Kaur M et al. Secure and energy efficient-based E-health care framework for green internet of things. IEEE Transactions on Green Communications and Networking. 2021;**5**(3):1223-1231

[184] Gadekallu TR et al. Blockchain-based attack detection on machine learning algorithms for IoT-based e-health applications. IEEE Internet of Things Magazine. 2021;**4**(3): 30-33

[185] Ayub MF et al. Lightweight authentication protocol for e-health clouds in IoT-based applications through

5G technology. Digital Communications and Networks. 2021;7(2):235-244

[186] Chehri A. Energy-efficient modified DCC-MAC protocol for IoT in e-health applications. Internet of things. 2021;**14**:100119

[187] Deebak B, Al-Turjman F. Secure-user sign-in authentication for IoT-based eHealth systems. Complex & Intelligent Systems. 2021:1-21

[188] Said AM et al. Efficient anomaly detection for smart hospital IoT systems. Sensors. 2021;**21**(4):1026

[189] Frikha T et al. Healthcare and fitness data management using the iot-based blockchain platform. Journal of healthcare. Journal of Healthcare Engineering. 2021;**2021**:12

[190] Huang C et al. A deep segmentation network of stent structs based on IoT for interventional cardiovascular diagnosis. IEEE Wireless Communications. 2021; **28**(3):36-43

[191] Hussain A et al. Security framework for IoT based real-time health applications. Electronics. 2021;**10**(6):719

[192] Alzahrani BA. Secure and efficient cloud-based IoT authenticated key agreement scheme for e-health wireless sensor networks. Arabian Journal for Science and Engineering. 2021;**46**(4): 3017-3032

[193] Iqbal N et al. A scheduling mechanism based on optimization using IoT-tasks orchestration for efficient patient health monitoring. Sensors. 2021; **21**(16):5430

[194] Amato F et al. A security and privacy validation methodology for e-health systems. ACM Transactions on Multimedia Computing, Communications, and Applications (TOMM). 2021;**17**(2s):1-22

[195] Luna-Perejón F, Domínguez-Morales M, Gutiérrez-Galán D, Civit-Balcells A. Low-power embedded system for gait classification using neural networks. Journal of Low Power Electronics and Applications. 2020; **10**(2):14

[196] Escobar-Linero E, Domnguez-Morales M, Sevillano JL. Worker's physical fatigue classification using neural networks. Expert Systems with Applications. 2022;**198**:116784

[197] Civit-Masot J, Luna-Perejón F, Corral JMR, Domínguez-Morales M, Morgado-Estévez A, Civit A. A study on the use of edge TPUs for eye fundus image segmentation. Engineering Applications of Artificial Intelligence. 2021;**104**:104384

Section 2

Novel Works in IoT

Chapter 4

An Effective Method for Secure Data Delivery in IoT

Mnar Alnaghes, Nickolas Falkner and Hong Shen

Abstract

The Internet of Things (IoT) has become very popular recently due to its important features that contribute to many aspects of our lives such as health and transportation. It consists of a vast number of different projects such as sensors, tags, actuators, and mobile devices, which can communicate and collaborate without human interactions. These devices carry small memory and low-energy battery, which affects their performance and lead to many issues. In this work, we are going to focus on the efficiency and security issues. We will propose a secure and efficient routing protocol for data delivery in order to improve its performance. The proposed technique will be evaluated in an implemented platform with appropriate case study. The expected outcome of this study will be a reference design and its practical implementation to support efficiency and security in IoT.

Keywords: IoT, data delivery, routing protocols, security efficiency

1. Introduction

Nowadays, many research efforts have been concentrated on the efficiency and security of IoT devices to raise the performance and the level of protection for IoT data and detect possible attacks. It is significant to understand the types of data delivery challenges in IoT. The challenges related to wireless sensor networks (WSN), cyber-physical systems (CPS), and machine-to-machine (M2M) continue to appear within the context of IoT since the basic components of IoT networks include WSNs, CPS, and M2M. One of the challenges is the difficulty in providing communications using infrastructure-based wireless systems because of the high cost of deploying and maintaining this infrastructure with the rapid growth of IoT users and devices [1]. Furthermore, the IoT system is mobile and dynamic; thus, its perimeters are not well-defined. It is also robustly heterogeneous concerning the devices, protocols, and communication medium.

The other concern is that IoT system is vulnerable to malicious cyber-attacks. One of these aggressive attacks is a distributed DoS (DDoS) attack, which intends to bring down a victim system by preventing legitimate devices of service from accessing it. DDoS attackers may also aim to gain unlimited access to the victim machines and cause more damage consequently. These attacks are made not to be significantly distinct from the usual behavior practiced by the system. One of the techniques that can be used to detect cyber-attacks is intrusion detection (ID). Yet, the advanced ID schemes utilizing machine learning techniques struggle to detect some of the cyber-attacks. These attacks are made not to be significantly distinct from the usual behavior practiced by the system. Therefore, there is a need for an anomaly-based IDS combined with artificial intelligence and machine learning due to its ability to classify and identify earlier hidden attacks. This kind of IDS will help in detecting multi-stage DDoS attacks. Current schemes in the development of ID investigate artificial intelligence and machine learning in academia and industry, such as artificial neural networks and fuzzy logic.

IoT advanced systems can achieve high performance with a human being's supervision for defining how to perform their duties. They also can automatically detect unusual patterns of web traffic with malicious activities and learn the patterns by themselves over time. Previous studies in the wireless network security area focused on ID based on a single hidden Markov model (HMM) and multi-class system classifier (MCSC) [2, 3]. Here, we study the potential applicability of the hierarchical hidden Markov model (HHMM) for intrusion detection in IoT systems in which the problem space can be several magnitudes higher than in wireless networks. And, we propose a probabilistic hierarchical hidden Markov model that reduces the high state-space without compromising classification accuracy. The proposed scheme shows better outcomes for detecting the DoS and DDoS attack patterns compared to the state-of-the-artwork.

The main contributions of the work are:

1. We propose a PHHMM model that translates high-dimensional IoT data to a discrete set of reliable data to be securely delivered with the ability to detect DDoS attacks.

2. We propose a method that learns and efficiently analyzes large amounts of data for classifying DDoS patterns in IoT traffic.

3. We conducted a performance comparison of our PHHMM with the baseline HHMM, Neural Network, and Naive Bayes models on the benchmark dataset from the CICIDS2019 database that contains 11 types of DoS and DDoS attacks collected over real-time for validating the proposed model.

The rest of this paper is organized as follows. Section II investigates the state of the art of some IoT data delivery used methods and presents the related work, followed by the background in Section III. Then, the model details are demonstrated in section IV and section V. Next, we show and discuss the experiments and simulation results in sections VI and VII. Finally, section V ends the paper, outlining some suggestions for future work.

2. Related work

2.1 Routing protocols in IoT

To design an efficient protocol in IoT networks is a risky task due to their characteristics. The efficient routing protocol has to respond to the changes that may happen in the topology as same as the bandwidth constraint. Most of the proposed protocols are only sub-optimal. Forster et al. [4] discuss three popular machine learning techniques on the communication layers in the WSNs. These algorithms are used in distributed environments to solve different problems such as ad hoc routing. They are categorized into three groups; reinforcement learning, supervised, and unsupervised. The aim is to find out a convergent mapping function that helps in prophesying the output results for any new input. Routing in IoT environments, as mentioned earlier, is associated with protocols in wireless sensors and ad hoc networks. One existing routing protocol for IoT networks is IP6 overpower personal area networks (6LoWPAN), which are used to route the data among non-IP sensors through networks with high processing capabilities. Its topology consists of a set of reduced function sensors that are linked to full function sensors [5]. It helps to support low cost, different length addresses, low bandwidth, different topologies, energy consumption, and lengthy sleep time. This protocol supports the multi-hop data delivery and reduces transmission overhead by providing header compression enclosing IPv6 long headers in the IEEE802.15.4 small packets [6]. Many of the real-world machine learning algorithms use both supervised and unsupervised learning as hybrid learning or semi-supervised learning to take advantage of the strengths of these main categories and minimize their cons [7]. Another standard protocol in IoT is the Routing Protocol for Low-Power and Lossy Networks (RPL) [8], which is a distance-vector protocol based on IPV6 that can prop lots of data-link protocols. It builds a destination-oriented directed acyclic graph (DODAG). It has only one path from each node to the root, and all the communications will be through that root. All nodes advertise themselves as the root by broadcasting a DODAG information object (DIO), and then the DODAG is gradually built. For the cognitive networks, as an extension of RPL, which is the Cognitive RPL Protocol (CORPL) is designed. It uses the DODAG topology generation. Constrained Application Protocol (CoAP) [6] is another IoT protocol that produces a lightweight RESTful (HTTP) interface to reduce overhead and power consumption. The next protocol is the Message Queue Telemetry Transport (MQTT), which was introduced for providing embedded connectivity between the party of middlewares and applications and the party of networks and communications. It is a publish/subscribe design that includes three parts: publishers, which are the sensors that connect to the broker to send their data, subscribers, which are the sensory data or applications, and the broker, which sends the data to the subscribers after classifying them in topics. Secure MQTT (SMQTT) [6] is an extension of MQTT to enhance its security features. It is encryption-based, where each message is encrypted and delivered to multiple nodes, which is common in IoT applications. For supporting a large range of IoT applications, ZigBee smart energy [6] is used. It has a wide star topology, peer-to-peer topology, or cluster-tree network topology. It also allows implementations with low memory and processing power. In addition, the Advanced Message Queuing Protocol (AMQP) [9] is designed for the financial industry. It is a publish/subscribe design built over TCP, but the broker here is divided into two main components: exchange and queues. The exchange receives publisher messages and distributes them to queues based on pre-defined conditions. Moreover, the long-term evolution advanced protocol (LTE-A) [9] is used for IoT applications in wireless networks. LTE-A design has a core network (CN) to control mobile devices, a radio access network (RAN) to establish data planes and control the wireless connections, and mobile nodes.

2.2 Intrusion detection systems in IoT

Because of the lack of training datasets, the current IoT intrusion detection systems are incapable of detecting the latest DoS and DDoS attacks [10], such as Network Time

Paper	HHMM	HMM	DDoS detection	Up-to-date dataset	Applicable for IoT
[1]	No	Yes	Yes	No	No
[2]	No	Yes	No	N/A	No
[3]	No	Yes	No	Yes	No
[4]	No	Yes	Yes	Yes	No
[5]	Yes	Yes	Yes	Yes	No
[6]	No	No	Yes	N/A	Yes
[7]	No	No	Yes	N/A	Yes

Table 1.
Intrusion detection schemes.

Protocol (NTP) attack, Network BIOS (NetBIOS) attack, UDP lag (delay). The authors in [11] proposed a hidden Markov model for predicting and detecting multi-stage attacks. Their work is not applicable for IoT systems as it fails on the high dimension state space since the incoming network traffic in IoT will have largely hidden states. The approach developed by [12] has a high detection rate as it identified most of the occurred attacks. However, they did not consider DDoS attacks. Authors in [2] proposed an anomaly detection module that uses Long Short-term memory for detecting both known and unknown attacks with a low false-positive rate. Their work shows high recognition rates. In [3], researchers discussed the multi-stage attack and its prediction. They proposed a multi-stage Naive Bayes model that can predict each stage of the multi-stage attack scenarios. However, schemes in [2, 3] are not suitable for predicting multiple attack intents in heteroecious environments. Besides, the authors in [13] propose a Hierarchical Hidden Markov Model (HHMM), which is an extension of the hidden Markov model (HMM), as the method for activity recognition. They analyzed the accuracy rate of their model with the Naive Bayes and HMM schemes. The comparison showed that the HHMM has the highest accuracy rate among others. However, they did not take into consideration the nature of IoT systems. The authors in [14] described routing-specific attacks in the IoT systems and concentrated on identifying the malicious node's location and neighborhood to inform the network administrator. In [15], the researchers proposed an ID scheme to detect flood attacks in IoT networks. Their proposed model identifies the attacks through the back-propagation neural network model. **Table 1** summarizes the properties of the major types of existing ID schemes.

3. Preliminaries

This section provides definitions for the used terms in this paper:

3.1 Distributed denial of service attacks (DDoS)

DDoS is one of the potential attacks in IoT where attackers coordinate the utility of many machines connected to the network to send an overwhelming amount of unwanted requests to a targeted server [16]. They try to disrupt the traffic of the server with a flood of unwanted requests. Besides, DDoS reaches effectiveness by using various compromised devices as the roots of attack traffic. The more hacked devices, the more damage is caused to the servers. Thus, the attacker examines remote machines for security gaps using some tools such as worms to find their vulnerabilities and inject them with the attack code. Then, these compromised machines become zombies, which the attacker uses to send malicious packets to the targeted victim. DDoS may yet cause a long-term memory consumption of the relaying nodes in IoT environments due to nodes' restricted resources. There are various DDoS attack types used to degrade the performance or availability of targeted services on the Internet. Some of these attacks are Botnet attacks, Spoof-packet flood attacks, Multi-Vector Attacks, and Misused Application Attacks. Besides, there are various schemes used to defend against DDoS attacks, which are under three categories; policy-based schemes, application-based schemes, and machine learning–based schemes. The policy-based defense scheme is placed in the switch to define the traffic that is allowed to be forwarded and the other ones are defined as malicious. It requires analyzing collected data samples of the network to classify malicious traffic. Numerous policy algorithms use different measurements such as standard deviation or measure the chi-square statistic of the sample to classify the packets as malicious or legitimate. Secondly, the application-based schemes handle and control packets in the network by the user interface layer. Finally, the machine Learning-based defense schemes deploy machine learning algorithms to investigate and classify the traffic to detect the DDoS attack.

3.2 Hierarchical hidden Markov model (HHMM)

The Hierarchical hidden Markov model (HHMM) is a multi-level stochastic process derived from the Hidden Markov model (HMM) by making each of the hidden states a self-contained autonomous probabilistic model. It is a statistical framework for modeling a sequence of observations. Each observation is emitted from a hidden state within the system by recursive activation. The basic idea of HHMM is that the upper-level states produce sequence states called "abstract" states [17]. And, the lower-level states produce single observations called "concrete" states [17]. The observations are governed by each of the sub-states (sub-HMMs). The process of recursive activations ends when reaching a state that produces output symbols like an HMM [17].

For estimating HHMM parameters, we define the generalized forward (α) and backward (β) probabilities as follows:

$$\alpha(i) = P\left(O, q_i^l | \lambda\right)$$
$$P(O|\lambda) = \sum \alpha_T(i)$$

where q_i^l is the number of sub-states of an abstract state.

$$\beta(j) = P\left(O | q_j^l, \lambda\right)$$

We also define the generalized horizontal (ξ) and vertical transitions (χ) as follows:

$$\xi\left(T, q_j^l, q^{l-1}\right) = P\left(i_t = q_i^l i_{t+1} = q_j^l | O, \lambda\right)$$
$$\chi\left(T, q_i^l, q^{l-1}\right) = P\left(i_t = q_i^{l-1}, i_{t+1} = q^l | O, \lambda\right)$$

The model is represented as $\lambda_{PHHMM} = \langle A^{q^l}, B^{q^l}, \pi^{q^l} \rangle$. And, the states of an HHMM are denoted by $Q^l = q_i^l$, where $l \in 1, 2, \ldots . L$, i is the state indexing, L is the output state, and l is the hierarchy indexing. It performs the following computation:

- A probability transition matrix ($A^{q^l} = a_{ij}^{q^l}$) is generated as the conditional probability of future traffic state is independent of the past states given the present state:

$$a_{ij}^{q^l} = P\left(q_{t+1}^{l+1} = S_j | q_t^{l+1} = S_i\right), \qquad 1 \leqslant i, j \leqslant N$$

$$a_{ij}^{q^l} \geqslant 0$$

$$\sum a_{ij}^{q^l} = 1$$

  where a_{ij} is a horizontal transition probability from state i to state j and all are sub-states of q^l.

  N is hidden states.

- An emission matrix ($B^{q^l} = b_{jh}^{q^l}$) for observation probabilities given the hidden traffic state is generated by:

$$b_{jh}^{q^l} = P\left(O_{h(t)} | q_t^l = S_j\right), \qquad 1 \leqslant i, j \leqslant N$$

$$1 \leqslant h \leqslant M$$

  where b_{jh} is observed probability in state j.

  M is observable states.

- An initial state distribution (π^{q^l}) is generated by:

$$\pi^{q^l} = \pi^{q^l}\left(q_i^{l+1}\right) = P\left(q_t\right) = s_1$$

4. Framework of the proposed model

In this section, we first explain the structure of the traditional HHMM model then we illustrate the framework of our proposed model.

4.1 Framework of the HHMM

Learning, decoding, and evaluating are the three principal HHMM objectives as described in [18]. Briefly, the techniques applied to achieve these objectives are as follows:

- **Learning:** Baum-Welch algorithm is used to create the object of the learning machine.

- **Decoding:** Viterbi algorithm is applied to define the most probable state path of hidden states that can be transitioned given an observation sequence (O) and the model parameters (λ).

- **Evaluation:** The forward and backward algorithms are used to determine the probability of an observed sequence.

This probabilistic hierarchical hidden Markov model should overcome the problem of the heterogeneity of IoT data. However, it suffers from high computational costs as the data increases in an exponential manner due to its used algorithms. Applying this scheme to IoT data undeviatingly will contribute to a problem of high state space. We, therefore, need to find a way to reduce the high state space without compromising the classification quality.

4.2 Framework of PHHMM

Our proposed model uses clustering and dimension reduction techniques to partition the massive incoming network traffic to overcome the problem of largely hidden states, before applying HHMM for classification. It follows the framework described in **Figure 1** and achieves the objectives [18] and techniques are applied as follows:

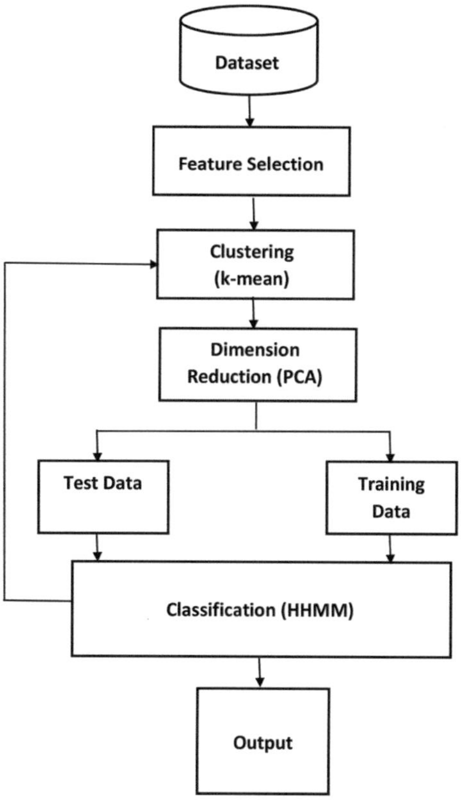

Figure 1.
The PHHMM detection model.

- **Estimate the model parameters:** Find the most probable parameter λ^* of the model by applying the Baum-Welch algorithm [19]: given the PHHMM structure and one or more observation sequences $\lambda = argmax_{\lambda^*} P(O_t|\lambda)$.

- **Learning: Calculate the probability of a sequence:** Derive the maximum likelihood estimate of the parameters of PHHMM given the set of output sequences by applying the Baum-Welch algorithm: given a PHHMM and its parameters, determine the likelihood $P(O|\lambda)$ of a sequence O to be generated by the model.

- **Decoding: Calculate the most probable state sequence:** Find the state sequences that best explain the observations by applying the Viterbi algorithm: given a PHHMM, its parameters, and an observation sequence, determine the single state sequence that is most likely to generate the observation sequence.

- **Evaluation: Detect DDoS attacks:** Evaluate the probability of the observed sequence and solve the detection problem to compute the probability of alert observations by applying the DDoS detection algorithm: given the probability of an observed attack alert sequence (O), detect DDoS attacks and predict future trails.

In our PHHMM model, applying dimension reduction techniques is a challenging step due to the lack of a standard approach for reducing the dimensionality of the observed IoT network traffic. It requires identifying the principal components and linear combinations of variables that describe the highest contrast in the massive data without compromising this data. Determining the principal components given a covariance matrix is computationally expensive as it claims the eigenvalue decomposition that requires the calculation of the covariance matrix. To overcome this challenge, we present an approach that avoids the direct computation of the covariance matrices but delivers the efficient subspace dimension. Our model applies the singular value decomposition (SVD) for calculating PCA to circumvent this expensive operation. The participating nodes in the algorithm use the PCA with the SVD learning mechanism to estimate principal components of the data traffic.

This work contributes to improving and resolving the common flaws in the application of HHMM in massive data by reducing the data dimensions based on traffic from both of the two streams being compared instead of depending only on some training data of normal traffic. Using only the most significant principal components, we could avoid the computation of the entire subspace. We can estimate a reduced number of principal components that are sufficiently effective in detecting malicious traffic. Our model allows the subsequent use of only the number of dimensions necessary at any given time.

5. Proposed model

This section explains the methods and techniques that we apply to the tasks in the tasks of the proposed model in **Figure 1**.

5.1 Data pre-processing

Collecting a large amount of attack traffic and normal traffic in a large real-time network is time and money-consuming. It needs significant resources, a diversity of normal IoT traffic, and a diversity of attack traffic. Instead, there are publicly available network traffic datasets, which can be used for this task. We analyze some of the available datasets based on the following:

- Real-time network traffic
- The most advanced DoS and DDoS attacks

The CICIDS2019 includes inbound and outbound traffic of the most advanced DoS and DDoS attacks. It contains lots of network flow-related characteristics and different types of current DoS and DDoS IoT attacks traffic collected over real-time networks.

5.2 Feature selection

The CICISD dataset has a huge number of attributes, selecting a subset of them is necessary for eliminating redundant and irrelevant ones. This helps in improving detection accuracy for DDoS attack detection. To better train our model for detecting the attack patterns, we have selected packet features that indicate DDoS attacks, which are useful for classification to distinguish between normal IoT traffic and DDoS IoT traffic:

- **Destination IP:** According to the research in [20], the IoT device communicates with several numbers of expected destinations that rarely get modified over time. Thus, if the device communicates with many separate destinations in a short time-stamp, it is considered an attack.
- **Packet size:** In [21], the authors declared that the size of a normal IoT packet size differs from 42 to 1434 bytes, while the size of a DDoS packet is smaller than 100 bytes. Thus, a sudden jump in the traffic with a fixed packet size of around 100 bytes represents a DDoS attack.
- **Packet Size Variance:** DDoS attack packets share the same size frequently whereas IoT packets' size differs from one to another.
- **Packet Time Interval:** The time interval between DDoS packets is close to zero while IoT packets travel in a regular time interval.

5.3 Data clustering

The k-means clustering is a method of vector quantization that aims to partition n observations into k clusters in which each observation belongs to the cluster with the nearest mean based on a similarity metric [20].

Let $S = f_1, f_2, \ldots \ldots f_n$ and $K = C_1, C_2, \ldots \ldots C_n$ then $C_i \neq \phi$, $C_i \cap C_j = \phi$, and $\cup_{i=1}^{K} C_i = S$, where $i, j = 1, 2, \ldots K$, and $i \neq j$.

The clustering method works as follows:

- **Input:** N features $\{f_1, f_2, \ldots, f_n\}$
- **Initialization:** Finding cluster centers $\mu_1, \ldots \mu_k$.

Randomly initialize K:

- Select K features as the primary cluster centers.
- **Reiteration:**
- Allocate each feature f_n to its closest cluster center $C_k = \{n : k = min_k f_n - \mu_k{}^2\}$.
- Determine the new center μ_k of the cluster C_k : $\mu_k = \frac{1}{|C_k|}\sum f_n$.
- Repeat until the cluster centers do not move anymore.

When deploying k-mean to cluster IoT traffic, we reduce the dimension data to quantize the data into single-dimensional data. We classify the traffic flow depending on similar characteristics and identify clusters of homogeneous traffic flows and define their borders. It needs to have high intra-cluster homogeneity and inter-cluster heterogeneity.

For determining the optimal (fitting) value of K, we use the elbow algorithm repeatedly applying different values of K and plotting their heterogeneity. When the curve begins to flatten, it reaches the optimal value of K.

5.4 Dimensionality reduction

After clustering the data, initially, we have n states $(S_1, S_2, \ldots, S_n)$, applying the dimension reduction technique results in a new set of m states $(s_1, s_2, \ldots, s_m)$ where $(m < n)$, $s_i = f_i(S_1, S_2, \ldots, S_n)$ and f_i represents a mapping function. Thus, its idea is to transform the massive data from a high-dimensional space, like IoT data, into a k-dimensional sub-space by partitioning the data-space into fully connected states. The low-dimensional form holds the top eigenvector v which has the meaningful features of the real data ideally close to its natural dimension [22]. The new set of features is extracted by some functional mapping. In this model, we considered the principal component analysis (PCA) and calculated it using the singular value decomposition (SVD) as the PCA presents a structure for reducing data dimensionality by outlining the maximum variation in the data.

To achieve dimension reduction by applying PCA, it requires placing the eigenvalues from the highest to lowest by their value. This ordering stores the elements in order of weight to the variance of the initial data matrix. This will allow us to drop the less important elements. Thus, we keep most of the information and lose a little noise. We can reduce the dimension of the original data. For instance, for any data of d dimensions, we only take the first r eigenvectors:

$$\frac{\sum \alpha_r}{\sum \alpha_d} = \frac{\alpha_1 + \alpha_2 + \ldots. + \alpha_r}{\alpha_1 + \alpha_2 + \ldots. + \alpha_d} \tag{1}$$

$$= \alpha_1, \alpha_2, \ldots., \alpha_r \tag{2}$$

Definition 1 [23]: For any matrix $Y = y_1, y_2, \ldots, y_n$ of the size $K \times d$ can be re-written as $Y = USV^T$ where, U is an orthonormal matrix of size $K \times r$, S is a diagonal matrix of size $r \times r$, V is a matrix of eigenvectors of size $r \times d$ (a column is an eigenvector) (see **Figure 2**).

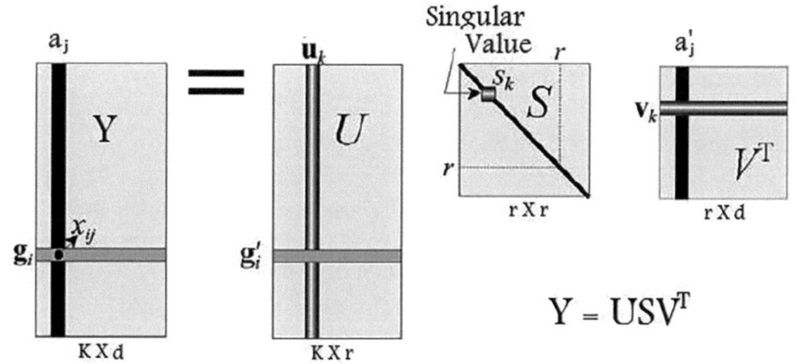

Figure 2.
Singular value decomposition of Y [23].

Assume that the data matrix Y is centered, i.e., the column means have been subtracted to be equal to zero. The covariance matrix (C) is calculated by:

$$C = \frac{X^T X}{K - 1} \tag{3}$$

Because the covariance matrix is symmetric, it can be diagonalized by:

$$C = VLV^T \tag{4}$$

where V is an eigenvectors matrix and L is a diagonal matrix with eigenvalues λ_i. The eigenvectors are called principal axes of the data, and, the data projections on the principal axes are called principal components [24]. After obtaining the singular value decomposition, C is defined by:

$$C = \frac{VSU^T USV^T}{K - 1} \tag{5}$$

$$= V \frac{S^2}{K - 1} V^T \tag{6}$$

As the result, the eigenvectors of C are the same as the matrix V (the right singular vectors of Y) and the eigenvalues of C can be defined from the singular values λ_i.

$$\lambda_i = \frac{s_i^2}{(K - 1)} \lambda_i \tag{7}$$

The principal components are defined by:

$$YV = USV^T V = US \tag{8}$$

In short, the PCA is calculated as follows:

- Obtain the maximum covariance in the data.
- Keep meaningful information only to reduce data size.

- Simplify the representation of the data:
- The variance of the data is maximized.
- Analyze the construction of the features and observations.

Based on Oja's algorithm for stochastic PCA optimization [25], the primary concept of our algorithm is to implement stochastic m updates by uniformly sampling the columns y_i at random, and reduce the variance of these updates.

$$X'_t = X_{t-1} + \eta y_{i_t} y_{i_t}^T X_{t-1} \tag{9}$$

$$X_t = \frac{1}{\|X'_t\|} X_t \tag{10}$$

We use the variance-reduced stochastic schemes for convex optimization [23] to reduce the stochastic variance. Let $B = \frac{1}{n} XX^T$; then the updates in each iteration can be rewritten of our algorithm as

$$X' = (I + \eta B)X_{t-1} + \eta \left(y_{i_t} y_{i_t}^T - B \right) \left(X_{t-1} - \overline{X}_{f-1} \right) \tag{11}$$

$$X_t = \frac{1}{\|X'_t\|} X_t \tag{12}$$

The algorithm is burst into periods f = 1, 2, 3,. .., wherein all period we do a single exact power iteration by computing $\overline{U}$. The steps to solve the problem are explained by the pseudo-code in Algorithm 1.

Algorithm 1: Dimensionality Reduction Algorithm.

Input Matrix $Y = (y_1, \ldots, y_n)$.
Output Matrix $\overline{X}_f$.
1: **Initialize** Orthonormal Matrix $\overline{X}_{k \times d}$.
2: **For** $f = 1, 2, 3 \ldots K$ **Do**
3: $\overline{U} = \frac{1}{n} \sum y_i \left(y_i{}^T \overline{X}_{f-1} \right)$
4: $\overline{X}_0 = \overline{X}_{f-1}$
5: **For** $t = 1, 2, 3, \ldots m$ **Do**
6: $B_{t-1} = VU^T$, where
USV^T is an decomposition of $X_{t-1}^T \overline{X}_{f-1}$
7: Set $i_t \in 1, 2, 3, \ldots n$ uniformaly at random
8: $X'_t = X_{t-1} + \eta \left(y_{i_t} \left(y_{i_t}^T X_{t-1} - y_{i_t}^T \overline{X}_{f-1} B_{t-1} \right) + \overline{U} B_{t-1} \right)$
9: $X_t = X'_t \left(X'^T{}_t X'_t \right)^{1/2}$
This is to ensure that W_t has orthonormal columns
10: **end for**
11: $\overline{X}_f = X_m$
12: **end for**

5.5 Hierarchical hidden Markov classification

Similar to HHMM [17], the PHHMM model uses the Baum-Welch algorithm to calculate the likelihood-maximizing parameters of the model given the observed data. It comprises four phases: the initial phase where the λ is randomly assumed if there is no prior knowledge, the forward phase where the forward variable is calculated recursively, the backward phase where the backward variable is calculated, and the update phase to update the parameters. Then, the model uses the Viterbi algorithm to find the most likely sequence of the hidden states given the observed data and the parameters. Finally, it uses the DDoS detection algorithm to detect the multistage attack based on the observed alert sequence, where, the standard HHMM focuses on a single category with a limited amount of features thus it is difficult to detect attack traffic that appears to be normal traffic.

The traditional HHMM constitutes multi-single states that are considered as self-contained probabilistic models [18]. However, due to the heterogeneity of IoT traffic, we design each state to have multiple separate lower HMM layers and one upper HMM layer, each lower state constitutes three levels:

- Learning: The observations of the first level train the L_1LHMM_i by the Baum-Welch algorithm to determine model parameters.

- Decoding: The observations of the second level trains the L_2LHMM_i by Viterbi algorithm to find the most likely sequence of hidden states using model parameters by the Viterbi algorithm.

- Evaluation: The observations of the third level trains the L_3LHMM_i and find the DDoS attacks sequence using the most probable sequence.

The model has one upper HMM state for predicting DDoS attacks that use the attack sequence from the lower states to learn new patterns of DDoS attacks by the DDoS detection algorithm. Thus, we can detect the multistage DDoS attacks in this extended mode, unlike the standard HHMM.

Model Training (Baum-Welch algorithm) Baum-Welch algorithm [19] is a recursive Expectation–Maximization method for estimating un-observed hidden parameters in an HHMM model. This algorithm facilitates the complex challenges of analytically applying maximum likelihood estimation. It trains the HHMMs to find the optimal λ. Starting with initialized values, the algorithm iteratively adjusts the parameters based on a set of observed feature vectors.

By this algorithm, for HHMM models $(\lambda_1^{q^l}, \lambda_2^{q^l}, \dots, \lambda_n^{q^l})$ and a given sequence of observations $\left(O^{q^l} = O_1^{q^l}, O_2^{q^l}, \dots, O_t^{q^l}\right)$, we choose $\lambda^{q^l} = \left(A^{q^l}, B^{q^l}, \pi^{q^l}\right)$ such that $P\left(O|\lambda_i^{q^l}\right), i = [1, n]$ is locally maximized.

Algorithm 2: The Baum–Welch algorithm.

Input Observation sequence O^{q^l}.
Output Re-estimated model parameters:

- The state transition matrix $a'^{q^l}_{ij}$, and.

- The Observation likelihood sequence $b'^{q^l}_{jh}$.

1: **Initialize** Observations (M), States (q_i), Threshold (Th).
2: Estimate $a_{ij}^{q^l}$, $b_{jh}^{q^l}$ using initialization techniques.
3: Calculate the expected probability $P(O|\lambda)$.
4: **reiterate.**
5: Calculate forward variable $\alpha\left(q_j^l\right)$:

$$\alpha_{t+1}^l\left(q_j^l\right) = b_j^{q^{l-1}}(O_{t+1}) \sum \alpha_t\left(q_j^{l+1}\right)\alpha_{ij}^{q_i^l}$$

6: Calculate backward variable $\beta\left(q_i^l\right)$:

$$\beta_t^l\left(q_i^l\right) = \sum b_j^{q^{l-1}}(O_{t+1})\, \alpha_{ij}^{q_i^{l-1}}\, \beta_{t+1}^d\left(q_j^l\right)$$

7: Calculate downward and upward variable $\varepsilon\left(q_i^l, q_j^l\right)$:

$$\varepsilon = \frac{\alpha_t\left(q_i^l\right)a_{ij}^{q^{l-1}}\beta_{t+1}\left(q_j^l\right)}{P(O|\lambda)}$$

8: Estimate the state probability $\gamma_h\left(q_i^l\right)$ and $\gamma_f\left(q_i^l\right)$:

$$\gamma_{h_t}\left(q_i^l\right) = \frac{\alpha_t\left(q_i^l\right)\beta_t\left(q_j^l\right)}{P(O|\lambda)} \qquad \gamma_{f_t}\left(q_i^l\right) = \frac{\alpha_t\left(q_i^l\right)\beta_t\left(q_i^l\right)}{P(O|\lambda)}$$

9: Compute the optimal state sequence $X(i,j)$:

$$X(i,j) = \frac{\alpha_t^l(i)a_{ij}^l\beta_{t+1}^l(j)b_j^l(O_{t+1})}{\sum\sum \alpha_t^l(i)a_{ij}^l\beta_{t+1}^l(j)b_j^l(O_{t+1})}$$

10: Estimate $a'^{q^l}_{ij}$ and $b'^{q^l}_{jh}$:

$$a'^{q^l}_{ij} = \frac{\sum \varepsilon\left(q_i^{l+1}, q_j^{l+1}\right)}{\sum \gamma_{h_t}\left(q_i^{l+1}\right)}$$

$$b'^{q^l}_{jh} = \frac{\sum \gamma_f\left(q_i^l\right) + \sum \gamma_h\left(q_i^l\right)}{\sum \gamma_f\left(q_i^l\right) + \sum \gamma_h\left(q_i^l\right)}$$

11: Calculate $P(O|\lambda)$ through the estimated parameters:

$$\varepsilon = P\left(O\middle|\lambda'^{q^l}\right) - P\left(O|\lambda^{q^l}\right)P\left(O\middle|\lambda^{q^l}\right) = P\left(O|\lambda'^{q^l}\right)a_{ij}^{q^l} = a'^{q^l}_{ij} \qquad b_{jh}^{q^l} = b'^{q^l}_{jh}$$

12: **until** $\varepsilon < Th$.
13: **return** $a'^{q^l}_{ij}, b'^{q^l}_{jh}$.

Model Decoding (Viterbi algorithm) Viterbi decoding algorithm [26] predicts the hidden traffic states. This algorithm only uses state-optimized joint likelihood for observation data and the underlying Markovian state sequence as the objective function for estimation. Opposed to the BW algorithm, it does not update all likely paths for all states in the HHMM.

Algorithm 3: Viterbi algorithm [26].

Input Re-estimated model parameters $a'^{q^l}_{ij}$, $b'^{q^l}_{jh}$ from BA algorithm,

Output Re-estimated state transition matrix $a'^{q^l}_{ij}$, and Alert observation likelihood sequence $b'^{q^l}_{jh}$.

1: **Initialize** Observations (M), States (q_i), Threshold (Th), a_{ij}, $b_j(v_k)$.
2: Obtain the model (λ) through expected probability $P(O|\lambda)$.

3: **reiterate.**

4: Split O^{q^l} into N states through Viterbi decoding.

5: $a'^{q^l}_{ij} = \frac{no.of\ states\ transitions\ fromitoj}{total\ no.of\ states\ transition\ fromi}$

6: $b'^{q^l}_{j}(v_k) = \frac{no.ofoccurrencesofobservationkinstatej}{totalno.ofobservationsinstatej}$

where, $b^l_j(v_k) = P\left(O^{q^l}_t = v_k | q^{q^l}_t = S_j\right)$.

7: Normalize row sums of $a'^{q^l}_{ij}$ and $b'^{q^l}_{j(}v_k)$ to unity so that all elements $\in$ [0,1].

8: Estimate λ'^l from O^{q^l}, $a'^{q^l}_{ij}$ and $b'^{q^l}_{j}(v_k)$.

9: $\lambda^d = \lambda'^{q^l}, a^{q^l}_{ij} = a'^{q^l}_{ij}, b^{q^l}_{jv_k} = b'^{q^l}_{j}(v_k)$.

10: **until** $|\lambda'^l - \lambda^l| > Th$.

11: **return** $a'^l_{ij}, b'^l_{jv_k}$.

Model Detection algorithm This algorithm uses prior knowledge to learn about the previous attack behavior and track the attack alerts. It gets the likelihood probability of the observation sequence $O^{q^d}_i$ then predicts the DDoS attack behavior based on the occurrence of the attack observation sequence in the previous algorithm.

Algorithm 4: Detection algorithm.

Input Alert observation sequence $O^{q^l}_i$, no. of iterations G, $\lambda^l_i = \left(A^{q^l}, B^{q^l}, \pi^{q^l}\right)$.

Output Attack alerts observation sequence $O'^{q^l}_i$.

1: **Initialize** $\alpha^{q^l_1}_0(j) = \pi^{q^l_1}_j b^{q^l_1}_j(O_0)$.

2: Calculate the probability of $O^{q^l}_1$:

$\alpha^{q^l}_j(1) = \pi^{q^l}_j b^{q^l}_j(O_1)$

3: Calculate the probability of observation sequence (O^{q^l}):

$\alpha^{q^l}_{t+1}(j) = |\sum \alpha^{q^l}_t(i) a^{q^l}_{ij} | b^{q^l}_j(O_{t+1})$

4: Calculate the likelihood sequence of the observable sequence (O) obtained by the model:

$P\left(O|\lambda^{q^l_i}\right) = \sum \alpha^{q^l_i}_t(j)$

5: If $(logP\left(O|\lambda^{q^l_i}\right) < -Th)$.

$alert_i = alert_i + 1$

6: **until** G is reached.

7: **return** *alert*.

---

6. Experimental setup

6.1 Datasets

In this section, the performance of the PHHMM based anomaly detection approach was tested on traffic combining DDoS attack data with normal data from the prepared dataset. We place the prepared dataset into our PHHMM model to identify DDoS attack intentions and predict the possible attacks. The performance of implementing our proposed model is obtained through MATLAB R2020b simulations. To remove duplicate alerts, we wrote a script for extracting necessary fields such as IP Addresses, Alert ID, Destination Port, Source Port, and timestamp from Snort IDS alerts.

6.2 Evaluation metrics

We analyze and evaluate the performance on the common metrics for IDS performance evaluation; Accuracy (the rate of true results including true negatives and true positives), Precision (positive predictive value), Sensitivity (true positive rate), Specificity (false positive rate), and False Negative Rate (error rate) [27], all in an average sense (see **Table 2**).

After generating the likely state sequences, we compare them to the known state sequences to define true positive (TP), false positive (FP), true negative (TN), and false-negative (FN) parameters [27]. The accuracy (ACC) is obtained by the following equation:

$$ACC = \frac{TP + TN}{TP + TN + FP + FN} \tag{13}$$

The precision (PR), the fraction of the total number of positive cases that are correctly identified as attacks to the total number of attacks, is obtained by the following equation:

$$PR = \frac{TP}{TP + FP} \tag{14}$$

Sensitivity (SN) or the true positive rate, the fraction of the total number of classified true positive that are accurately identified as attacks to the total number of positive cases, is calculated by the following equation:

$$SN = \frac{TP}{TP + FN} \tag{15}$$

We use $F_{measure}$ to evaluate the model's overall accuracy considering both precision and sensitivity. Having a good F-measure value indicates that the model has low false positives and false negatives, which means that it correctly identifies attacks. It is calculated by the following equation:

$$F_{measure} = 2 \times \frac{PR \times SN}{PR + SN} \tag{16}$$

		Predicted class	
		Attack	**Non-attack**
Actual class	Attack	TP	FN
	Non-attack	FP	TN

Table 2.
Two-Class Case Confusion Matrix.

The following equation is used to identify the error rate (ER) for false negative predictions:

$$ER = \frac{FP + FN}{TP + TN + FP + FN} \quad (17)$$

7. Evaluation results

Compared to the original system, our model constructs an equivalent system with a minimal number of constraints over real-valued variables consisting of bounds on variations. This helps in reducing the high state space and improving the classification accuracy and time complexity as well.

Models (Training 80%/ Testing 20%)	*ACC*	*PR*	*SN*	$F_{Measure}$	*ER*
Neural network	0.92	0.90	0.94	0.92	0.05
Naive Bayes	0.45	0.68	0.56	0.61	0.06
HHMM	0.975	0.92	0.979	0.94	0.03
PHHMM	0.989	0.979	0.99	0.985	0.02

Table 3.
Experiment results summary I.

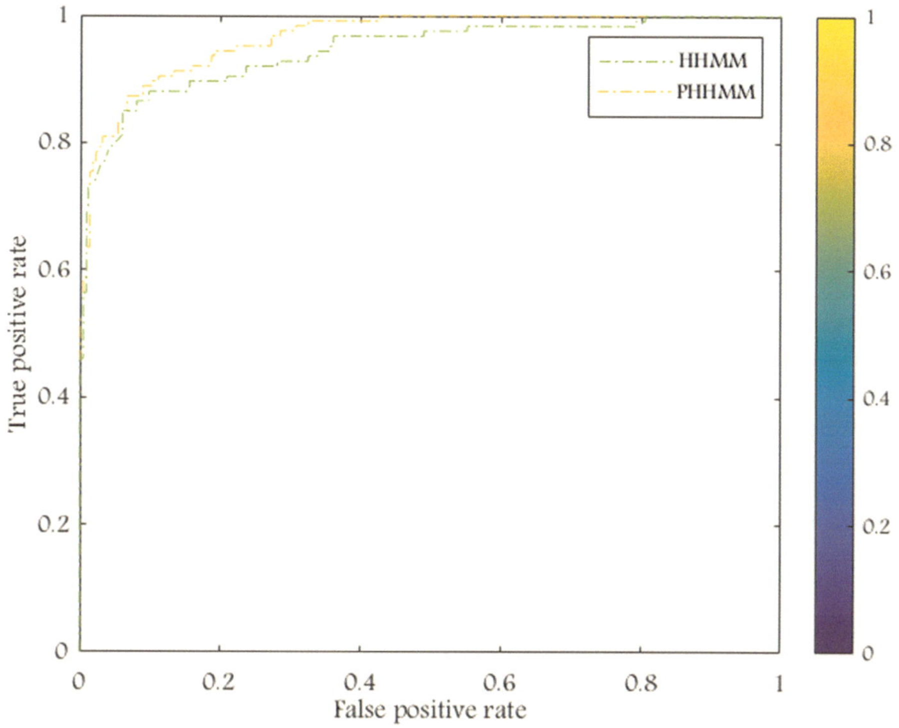

Figure 3.
Roc curves of models performance.

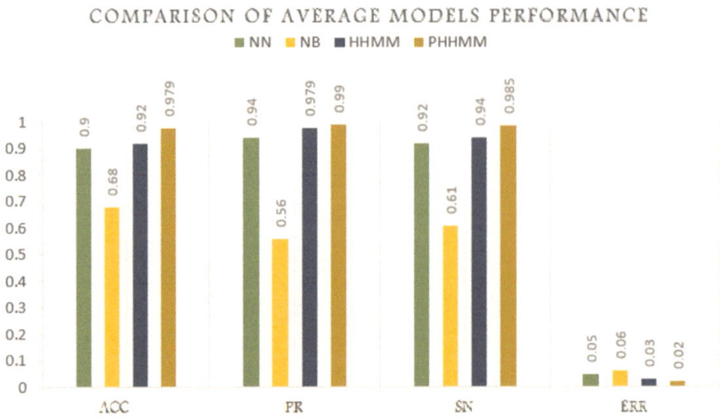

Figure 4.
Comparison of average models performance.

7.1 Classification accuracy

The above results show that our proposed model obtains satisfactory results with regard to attack detection rate. The proposed model has 98.9% accuracy and 97.9% precision. **Table 3** reviews the results of the performance of our model compared to the neural network (NN), Naive Bayes (NB), and HHMM classification algorithms.

We perform the tests using different window sizes to understand their influence on the detection. It shows that increasing the size of the window results in better accuracy.

Figure 3 shows the ROC curves for the performance of our proposed model, compared to the HHMM, NN, and NB models. ROC curves help identify the balance between the true-positive rate and the false-positive rate for all possible thresholds. It illustrates the model's strength to differentiate between attack and non-attack classes (see **Figure 4**).

7.2 Efficiency

Computation time is not associated instantly with classification; however, it describes the training time taken by the model. **Table 4** shows that our model has a lower computation time compared to the HHMM. In [28], The time complexity of calculating the probability of a sequence and estimating the HHMM parameters as the model depends on the length of the observation equals $O(ST^3)$, where S is the number of states, and, T is the granted transactions number at each level. Meanwhile, the PHHMM time complexity equals $O(S\log(S))$ based on our calculations. **Figure 5** shows the time complexity of HHMM vs. PHHMM.

Models	Computation time (sec)
HHMM	8.3
PHHMM	5.6

Table 4.
Comparison of computation time.

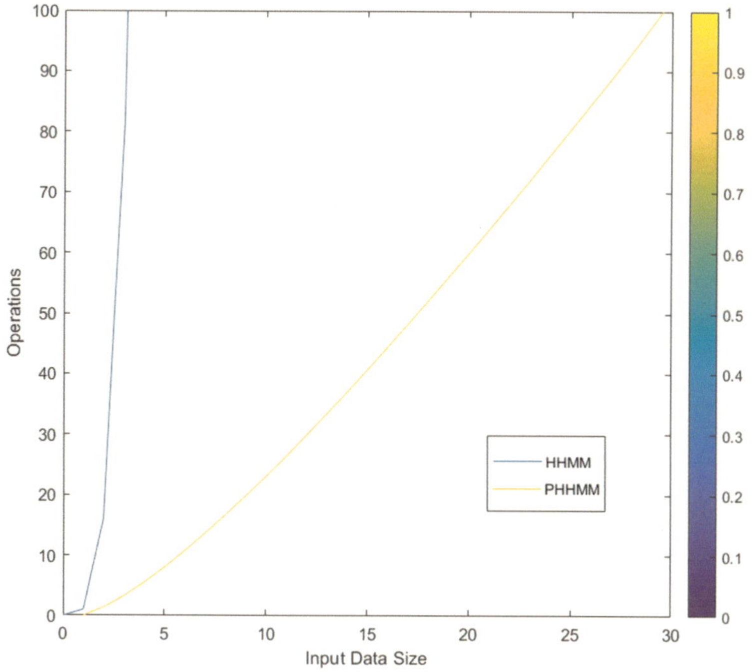

Figure 5.
Time complexity of HHMM vs. PHHMM algorithms.

8. Conclusion

In this work, we propose a probabilistic hierarchical hidden Markov model (PHHMM) applied for IoT intrusion detection which is more efficient than the existing HHMM without compromising classification accuracy. The main idea of our model is to reduce the huge problem state space of IoT traffic through dimensionality reduction by PCA and SVD. The proposed model is tested on the CICISD2019 dataset to detect and predict DDoS attacks. We evaluated our model on major performance metrics including Accuracy, Precision, Sensitivity, and False Negative Rate, Specificity and shows that our scheme has better detection accuracy and low error rates compared to Naive Bayes and neural network classification algorithms. It shows that PHHMM achieves a comparable accuracy as HHMM, better than NB and NN, and better efficiency than HHMM.

Acknowledgements

We would like to thank the Ministry of Higher Education in Saudi Arabia and the University of Adelaide for partially supporting this research.

Author details

Mnar Alnaghes[1*], Nickolas Falkner[1] and Hong Shen[2]

1 University of Adelaide, Adelaide, SA, Australia

2 Faculty of Sciences, School of Computer Science, University of Adelaide, Engineering, and Technology, Australia

*Address all correspondence to: mnar.alnaghes@adelaide.edu.au

References

[1] Zhang J, Hu P, Xie F, Long J, He A. An energy efficient and reliable In-network data aggregation scheme for WSN. IEEE. 2018

[2] Chadza T, Kyriakopoulos K, Lambotharan S. Analysis of hidden Markov model learning algorithms for the detection and prediction of multi-stage network attacks. FGCS. 2020

[3] Ingale S, Paraye M, Ambawade D. Enhancing multi-step attack prediction using hidden Markov model and naive Bayes. ICESC. 2020

[4] Forster A, Murphy AL. Machine Learning across the WSN Layers in Emerging Communications for Wireless Sensor Networks. Rijeka: In Tech; 2011

[5] Chelloug S. Energy-efficient content-based routing in internet of things. Journal of Computer and Communications. 2015

[6] Salman T, Jain R. Networking Protocols and Standards for Internet of Things. John Wiley and Sons, Inc.; 2017

[7] Iqbal M, Bayoumi M. Secure end-to-end key establishment protocol for resource-constrained health care sensors in the context of IoT. In: Proceeding of the HPCS. 2016

[8] Korte K, Sehgal A, Schönwälder J. A Study of the RPL Repair Process using ContikiRPL. In: Proceedings of the 6th International Conference on Autonomous Infrastructure, Management, and Security. 2012

[9] Alam S, Siddiqui S. Security & Privacy Threats, attacks and countermeasures in internet of things. International Journal of Network Security and Its Applications. 2019

[10] Beslin Pajila PJ, Golden Julie E. Detection of DDoS Attack Using SDN in IoT: A Survey. Springer; 2020

[11] Brogi G, Bernardino E. Hidden Markov models for advanced persistent threats. IJSN. 2019

[12] J. Koo, and S. Cho,"Effective intrusion type identification with edit distance for hmm-based anomaly detection system", in Pattern Recognition and Machine Intelligence. Springer, 2005.

[13] Suratkar S, Kazi F, Gaikwad R, Shete A, Kabra R, Khirsagar S. Multi hidden Markov models for improved anomaly detection using system call analysis. IBSSC. 2019

[14] Bhosale S, Sonavane S. A real-time intrusion detection system for wormhole attack in the RPL based internet of things. Procedia Manufacturing. 2019

[15] Pham T, Yeo C, Yanai N, Fujiwara T. Detecting flooding attack and accommodating burst traffic in delay-tolerant networks. IEEE Transactions on Vehicular Technology. 2018

[16] Nazih W, Hifny Y, Elkilani WS, Dhahri H, Abdelkader T. Countering DDoS attacks in SIP based VoIP networks using recurrent neural networks. Sensors. 2020;**20**

[17] Kong W, Dong D, Hill D. A hierarchical hidden Markov model framework for home appliance modeling. IEEE Transactions on Smart Grid. 2016

[18] Fine S, Singer Y, Tishby N. The hierarchical hidden Markov model: Analysis and applications. Machine Learning. 1998;**32**

[19] Chadzaab T, Kyriakopoulosa K, Lambotharan S. Analysis of hidden Markov model learning algorithms for the detection and prediction of multi-stage network attacks. Future Generation Computer System. 2020

[20] Kumar S, Raza Z. A K-means clustering based message forwarding model for the internet of things (IoT). Confluence. 2018

[21] Otoum Y, Liu D, Nayak A. DL-IDS: A deep learning-based intrusion detection framework for securing IoT. Transactions on Emerging Telecommunications Technologies. 2019

[22] Pal S, Bandyopadhyay S, Biswas S. Pattern recognition and machine intelligence. PReMI. 2005

[23] Tenenbaum J, Silva V, Langford J. A global geometric framework for nonlinear dimensionality reduction. Science. 2000

[24] Smith L. A tutorial on Principal Components Analysis. 2002.

[25] Johnson R, Zhang T. Accelerating stochastic gradient descent using predictive variance reduction. NIPS. 2013

[26] Bongiovanni W, Guelfi A, Pontes E, Silva A, Zhou F, Kofuji S. Viterbi algorithm for detecting DDoS attacks. LCN. 2015

[27] Lawal M, Shaikh R, Hassan S. An anomaly mitigation framework for IoT using fog computing. Electronics. 2020

[28] Chunfu J, Feng Y. An intrusion detection method based on hierarchical hidden Markov models. Wuhan University Journal of Natural Sciences. 2007

Chapter 5

Autonomous Update of a Dataset for Anomaly Detection Services in Elderly Care Smart House

Linos Nchena and Martin Tomášek

Abstract

This work proposes a smart system that could be useful in the delivery of elderly care services. Elderly care is a set of services that are provided to senior citizens to help them have a more comfortable and independent life which would not be possible without these services. This proposed system is unique in that it combines the detection algorithm with the automatic update of the dataset. It also uses a heuristic mechanism to reduce false detections. This is on the premise that the AI effort is good, but it could be made better with the inclusion of heuristics. Fall detection accuracy is initially solved by the first classifier, then another classifier evaluates the result with inferences before evoking an alarm. It checks the location of the subject to use in its inferences. Hence the smart house design consists of two machine learning systems. One system performs human activity classification while the other performs fall occurrence detection. Of the eight different classification methods utilized, XGBoost was most accurate with an average of 97.65% during training. A customized dataset is then generated with newly labeled data hence improving system performance.

Keywords: artificial intelligence, machine learning, human activity recognition, activity of daily living, fall detection, fall prevention

1. Introduction

According to the United Nations [1], in 2025, there would be a total of about 1.2 billion people over the age of 60. By 2050 this number would increase to 2 billion with 80% of them living in developing countries. The population growth has increased for older persons than for the rest of the population. In 1950 the total number of people over 60 years old was 8 percent. By 2007, this percentage had grown to 11 percent. By 2050, this number is projected to be at 22 percent. This kind of population growth comes with various challenges of its own [2].

This increase in the elderly population means more people would require assistive living care than they were before. Thus, Smart houses could play a key role in attending to this huge elderly population that needs assistive living care. A smart house is a special type of house that has automated services delivered by that house. Smart houses are of diverse types based on their purpose. Some of the common types of smart houses include (a) healthcare-oriented, (b) entertainment-oriented (c) security-oriented, and (d) energy-efficiency-oriented smart houses. In this research, we present a smart house model that is based on specific requirements for elderly citizens. With an emphasis on the needs of assistive technologies (AT), we shall recommend a smart house design. The design shall satisfy three major requirements which are vital for senior citizens. These three requirements are; (a) ATs services, (a) privacy requirements, and (c) security services. We are aiming to develop a smart house system which consists of several monitoring services This system should also enable modularization and allow easier replacement of components.

With the vast improvements in medicine and quality of living, many people are now living longer lives than previously was possible. This has been a result of vast investment in research that will improve quality of life. In trying to improve the quality of life, several researchers have attempted to provide a solution to care for senior citizens [3]. These solutions need enhancement to build completely novel solutions to deal with the growing demand for senior citizen care soon. The purpose of this work is to explore how this problem can be controlled using assistive technologies. To help assist with this issue we shall have to perform three experiments. We need to create a system that can be able to tell when an anomaly has occurred in the senior's smart house. This requires knowing what is and what is not an anomaly. We have data that is recorded from the activity in the smart house using a conventional sensor such as those in mobile phone sensors or smartwatches. We shall require label data to determine if the data tread is normal or abnormal. This can be achieved by using the labeling used in the previous experiment dataset. Several publicly available datasets exist. Among the common dataset include Sisfall, MobiAct, Ucihar, Unifall, and Unimab datasets [4–6]. These datasets provide acceptable benchmarks to determine the classification of ADLs and falls. These could be used in the classification of data, or to assess a system's accuracy.

In a smart house, assistive technologies are installed to detect abnormalities in human activity or environmental parameters. This is achieved using several methods. Three of the common methods are threshold, heuristics, and machine learning [7]. Threshold systems used specified rules in which a dataset is evaluated on those rules. Based on these rules a censored dataset can be labeled as a fall or not a fall. Using machine learning a different approach is used. A network is created which has node relationships that can be able to determine whether an activity is a fall or not has occurred. ML works similarly to a Blackbox solution as the rules are not logically deductible easily in the network c. Several machine learning classification algorithms exist in the labeling of subject data. Nine of the common classical ML algorithms are k-Means, Linear discriminant analysis (LDA), Naïve-Bayes, K-nearest neighbor (KNN), Vector support machine (SVM), Artificial neural network (ANN), Random Forest, and Decision trees [8].

Moreover, when collecting personal data, security and privacy should be considered. For example, cameras might capture more private information than smartwatches, some of this information can violate privacy. When using the toilet, the activity is not appropriate to record on camera while a smartwatch record of toilet activity might be more acceptable. Hence the choice of sensor method is especially important in developing this system. However, take note that it may be easy for detecting activity with a camera than with smartwatches. Therefore, a compromise needs to be considered in such cases.

The rest of this article is organized as follows; In Section 2, we describe some of the previous works related to ours by other researchers. In Section 3 we describe the methodology of what would be performed and how it would be performed. The section describes the flow of the algorithm and the dataset used. We then discuss the results of the experimental works in Section 4. In Section 5 we discuss issues that are related to our results and the future directions of this research. In Section 6 we present our findings and conclusion from this research and what is next.

2. Related research

Several people have attempted to solve this problem. The following are the most interesting of the research works which are of interest to this article's research aims.

According to the article [3], a solution is proposed where a group of agents work together to sense communicate and interpret sensor data. The agents are seven types which consist of communication, sensor, refining, reconstruction, interpretation, prevention, and cognitive agents. The agents are separated into two groups. The group of agents each processes the sensor data and then aggregates the result to form a concrete decision. The first group of agents was prediction activities in the smart house. The performance was tabulated as 72.00% for machine learning, 88.00% for expert-knowledge agents, and 91.33% for meta-prediction agents. The last group of agents was prevention agents. This was a simulation and achieved 100% accuracy. However, the last group was not real-life but simulative experiment.

The researcher in the article [9] presents a monitoring system for senior citizens. If an anomaly is identified, then the system will send an alarm to a caregiver. Some activities monitored include waking up in the morning, preparing food, having breakfast, reading, working on a computer, having lunch, napping, or reading. An example of an anomaly is where the system detects that she woke up and starts walking around the house at 2 pm. This is an anomaly because this time is an awkward time for walking around the house. An alarm is evoked as this is not part of the normal schedule. A mock apartment was designed for use in this experiment.

The study in the article [10], graphical presents a comparison and heuristic technique was utilized in detecting falls. A publicly available suit GBAD test suite was defined. It selects the best subgraph or pattern and then compares it with the sensor data. Each graph is compared to the full graph using the formula. An abnormality can be identified from the graph in the suit presented.

2.1 Datasets with classification labels

In article [11] a dataset (Sisfall), and a fall detection algorithm are presented. The algorithms have five stages which are included in this order; sensor data, pre-process data feature extraction fall detection, and finally, call for help when a fall is detected. Four algorithms were used, and these are decision tree (DT), Logistic regression, k-nearest neighbor (KNN), and support vector machine. The dataset that was used is called Sisfall. Sisfall contains 15 types of falls and 19 types of ADL. These were performed by 23 young people aged 19 to 30 years old and 15 elderly people aged 60 to 75 years old. The recording frequency was 200 HZ for sampling. The sensors used were two accelerometers and one gyroscope. The accuracy was derived from the relationship between true positive (TP), true negatives (TN), false positive (FP), and false negatives (FN). Accuracy is defined as in Eq. (1) below.

$$Accuracy = \frac{TP + TN}{TP + FN + TN + FP} x100 \quad (1)$$

The accuracies recorded were DT at 99.02%, LR at 99.38%, KNN at 99.91%, and SVM at 99.98%. The most accurate results are the SVM classifier. SVM was found to have performed not only better in this experiment but also better than selected previous benchmark works of previous performances.

In article [12], a dataset (MobiFall) is presented. An experiment was conducted to standardize a dataset that can be used to determine if there is or no fall in a sensor. Datasets are used in machine learning to benchmark and identify specified activities. They compared two fall detection systems. One threshold-based system and another machine learning-based system. The machine learning system had a higher accuracy level. In this dataset, four kinds of falls were studied. Forward-lying, front-knees-lying, sideward-lying, and back-sitting-chair. Apart from fall detections, ADLs were also studied. These were nine which include standing, walking, jogging, jumping, starting up, stair down, sitting in a chair, car-step in, car step out. The sensor data used was from three types of sensors, the accelerometer, gyroscope, and orientation signals. The Size of displacement defined by the slope (SL) was measured using the formula SL given as in Eq. (2) below.

$$SL = \sqrt{(max_x - min_x)^2 + (max_y - min_y)^2 + (max_z - min_z)^2} \quad (2)$$

Where the X stands for X-axis displacement, Y is the Y-axis displacement and Z is the Z-axis displacement.

Accuracy in fall detection was at 98% and in fall classification was at 68% using the 10-fold cross-validation. However, in another method where two-thirds are for training and one-third for testing, the accuracy was fall detection at 98.74 and fall classification at 68%. The dataset used is called MobiFall. This dataset is publicly available at Hellenic Mediterranean University (HMU) in Crete, Greece.

In article [13] a dataset (Ucihar) and six classifiers are used to detect falls. These involve distinguishing falls from ADLs. The six classifiers are the k-nearest neighbor (k-NN), least squares method (LSM), support vector machines (SVM), Bayesian decision making (BDM), artificial neural networks (ANNs), and dynamic time warping (DTW), Fourteen people performed the experiment for data acquisition. The trial had 20 falls and 16 ADLs. The formula used to determine the total acceleration is given in Eq. (3) below.

$$A_T = \sqrt{(A_x)^2 + (A_y)^2 + (A_z)^2} \quad (3)$$

where A_x is acceleration in the x-axis, A_y is in the y-axis, and A_x is in the z-axis.

A database was created containing fall activities and ADLs. All the six algorithms performed at around 95% with K-NN and LSM being the most accurate. The researcher suggests using these two algorithms for live data stream detections. This dataset is accessible at the University of Irvine Machine Learning Repository.

2.2 Types and placement of sensors

For the recording and collection of data, specific sensors are acquired. Several types of sensors exist. Wearable sensors are sensors that can be placed on the body of interest. Environmental sensors are sensors that are not embedded in a body of interest. Data collection should be done at a regular frequency. However, we shall assume that the sensors wherever they would be placed would collect the same type of data at the same frequency. This is on the pretext that, if the location of the sensor is changed the frequency and the quality of data might as well change.

In article [14], two types of sensors are defined. Vision-based and sensor-based. The vision sensors use cameras of diverse types. These sensors however are not very much acceptable to the intended beneficiaries of the system. Another type is sensor-based, which includes wearables ambient sensors, and sensors on an object. This is the most used type as it is considered less intrusive and is more acceptable by potential beneficiaries. An accelerometer and gyroscope are two examples of wearable sensors. In article [15] three sensory systems are defined. These are wearables, vision, and ambient sensors which are a combination of visual with sound and location sensors. The visual sensors become more data accessing with the addition of location sensors and sound sensors. The human voice can also be used as input. A representation of the three types of sensors is shown below in **Figure 1**.

In article [16] a system of sensors is proposed. Environmental and wearables sensors are combined to identify the location and the motion of the subject person. Each of these has achieved a significant level of activity classification accuracy. It is declared that using only an accelerometer the accuracy is 54.19%, while if you combined an accelerometer with environmental sensors the accuracy is 97.42%.

In research [17] the authors say the approach is intrusive as it requires active dressing, and it could be monitoring specific categories and not extensive categories of activities. It also might require the continuous wearing of the sensors throughout the day to enable data sensing. Therefore, researchers argue that environmental sensors are much better than wearable sensors. They provide experiments using a vision-based sensor. The activities involved include sitting, standing, walking, sleeping, getting assistance, and using the bedside commode and background. The camera has two kinds of data frames. A thermal defame and depth frame. They adjust each frame to its best resolution for better results. Although the authors acknowledge the intrusive and costly nature of vision sensors, they insist that more can be done with visual sensors to be used in the detection of activities in houses of seniors.

In article [18] a voice is used as input for a smart house. The smart house must interpret this voice according to the training and evoke some devices to perform a particular action. The purpose of the article was to provide voice input with a secure connection to a smart house with IoT Network. In article [19] the author mentions that training is a difficult part of a detection system. This is because the training dataset can become obsolete when the individual trained on it changes their usual pattern. The mentioned in cases when senior develops diabetics or drift concept. With advanced age, this leads to body deterioration which results in changes in gait

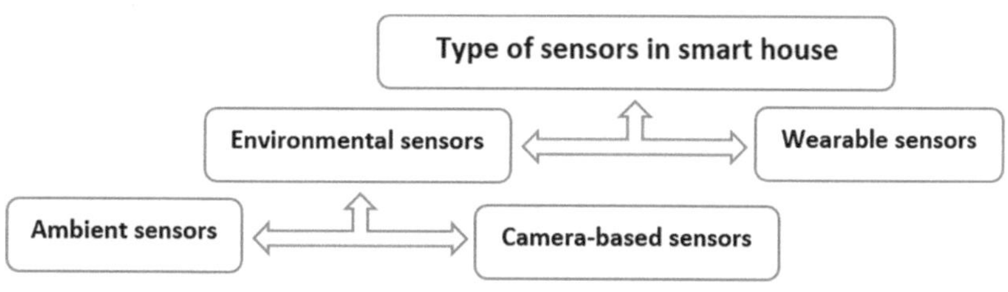

Figure 1.
Types of sensors in the system design.

No	Cited	Type of sensors used in experiment	Classifications models utilized	Dataset	Accuracy
1	3	Accelerometers	Random Forest, Hidden Markov Model, Support vector machine, Decision Tree (C45)	Lab Simulated	72.00%
2	9	Wearable sensors, Vicon system (PIR), camera	Dynamic Bayesian Network	Lab Simulated	99.00%
3	10	Infrared sensors, thermometers, object identifies, burning sensors, door sensors	Graph-Based Anomaly Detection (GBAD tool)	Kyoto dataset (CASAS-400)	n/a
4	11	Accelerometers, and a Gyroscope	Support vector machine, Decision Tree, Logistic Regression, K-Nearest Neighbor	Sisfall	99.98%
5	12	Accelerometer, Gyroscope and Orientation signal	K-Nearest Neighbor, Naïve Bayes, Decision tree-J48, Random Forest, Support vector machine	MobilFall	98.41%
6	13	Accelerometer, gyroscope, and magnetometer/campus	K-Nearest Neighbor, Support vector machine, BDM, Decision Tree, ANN	University of Irvine Machine	95.00%
7	15	Accelerometers and a Gyroscope	CNN-Long short-term memory (LSTM)	UMA Dataset	86.63%
8	16	Thermometer, Humidity sensor, Light sensors, Orientation and Motion sensors	Random Forest, Naive Bayes, Bayes Net, Logistic Regression, MLP, Radial basis function (RBF), Decision Table, Decision Trees(J48), Random Tree	CREST testbed,	99.13%
9	17	Depth sensor Camera, thermal sensor camera	Convolutional Neural Network - (ResNet-34 architecture)	Experimental testbed	95.80%
10	19	3D Accelerometer and Gyroscope	Multi-layer perceptron (MLP)	MobiAct	98.75%

Table 1.
Listing of sensors and classifier methods in selected previous research.

characteristics of seniors [20]. A summary of sensors and classifiers from related works is shown below in **Table 1**.

3. Research methodology

In this section, we shall describe the tools and the methods we shall use. We shall discuss the sensors used, the datasets used, the algorithms utilized and the floor plan of the apartment that will be used for the senior's smart house.

3.1 Implementation procedure, software, and hardware

Since the research is about seniors, the dataset had to be slashed to only utilize data that applies to seniors. The age of participants was the main factor used to extract records. Some activities were removed as we considered them not necessary for seniors. Based on this In the Sisfall of 56,786 records, only 14,000 have seniors. In the MobiFall dataset on the 62,259 records, only 16,598 have seniors. In the Ucihar dataset on the 7352 records, only 2018 has seniors. After extracting these records, seven classifier algorithms were used to train the model on all three datasets and then validate the models.

The classification was for two separate tasks. The first task was fall detection. The MobiAct and Sisfall were used in this task. Each dataset has its subsection used in the experiment. The performances of the algorithms were compared to the dataset sub-class from these two datasets. The best algorithm was identified. The second task was to identify if the ADLs were Laying or not. In this task, only the Ucihar dataset was used. The eight classifiers were again used to detect if Laying was the activity performed or not. This task was then linked to some other external tasks that tell the location where Laying was occurring if Laying was identified as the current ADLs.

The experiment was conducted using Python 3 software package. The computer system had the following specification: Processor: Intel(R) Core (TM) i7-4510U, CPU: 2.00GHz, 2 cores, RAM Memory: 4 GB DDR3 1600 MHz, OS: Windows 64 bits.

3.2 Sensors

The sensors we shall use in collecting data are a gyroscope and accelerometer for the XYZ plane. Nowadays These are readily available in mobile phones and smartwatch devices, which are usable for human activity recognition. The sensor using cameras and environment sensors has the advantage that they do not require excessive preparation and arrangement for senior citizens. However, the sense that the senior is under surveillance might be a discomforting feeling for most seniors. Therefore, this discomfort has made us decide to use an accelerometer and gyroscope other than cameras. Therefore, in this work gyroscope and accelerometer which are embedded in smart are utilized. This sensor data can be analyzed and processed in separate locations. The data we collected from the sensors is not labeled. Therefore, to enable labeling we have mapped our data to the data labeled by previous researchers. The labeled data will be obtained from publicly available data sets.

3.3 Datasets

Several datasets exist for human activity recognition (HAR) and fall classification. These can be used to classify test data as either a fall or an ADL. In our research, we shall use the publicly available datasets; SisFall, MobiAct, and Ucihar dataset.

In research [11, 21] the Sisfall dataset is generated in which most of the subjects are between the ages of 20 and 47 years. The primary research was to create a prototype dataset for fall detection. The second dataset is MobiAct in research [12, 22]. In the MobiAct dataset, the subjects' ages are from 40 to 47 also doing both the falls and the ADL. This Sisfall and the MobiAct dataset are used primarily for the exploration of fall detection. The third dataset is the Ucihar [13, 23]. This dataset does explore the identification of ADL. There is no fall detection in Ucihar. We shall use this in ADL identification where appropriate as shall require a labeled dataset that could define ADL. In our research we focus on senior citizens hence we prefer data for people about 60-year-old and above. However, this data is not readily available hence shall show infer it in various datasets. Below is **Table 2** which shows the composition of the dataset to be used in the experiment.

No.	Dataset	Human activity recognition samples	Fall detection samples	Total number of data samples
1	Sisfall	44,795	11,991	56,786
		79%	21%	100%
2	MobiAct	50,188	12,071	62,259
		81%	19%	100%
3	UCIHAR	7352	0	7352
		100%	0%	100%
4	Total	102,335	24,062	126,397
		81%	19%	100%

Table 2.
Composition of the selected datasets used in the experiment.

The experimental devices must be held compatible with mobile devices that were sensing the data. For fall detection we shall classify falls as dangerous activities. Therefore, we should need to send a warning message if a fall occurs unlike when an ADL occurs. The data has been labeled by the above public libraries (Sisfall, MobiAct, and Ucihar). We shall use the accelerometer, which records the speed of objects. And using this speed we could tell the presence or absence of a fall. The gyroscope would be used for rotational movements which is another parameter in the detection of a fall by the senior citizen. The accuracy is the efficiency of the system. We shall use selected sections from the three sample datasets, which are more appropriate per our requirements.

After preprocessing the raw data, a training and test dataset is derived which has a smaller number of records. The total data fields are nine both in MobiAct and Sisfall as shown in **Figures 2** and **3** respectively.

Ucihar is for the activity classification classifier. Unlike the above dataset, the original dataset has a total of 548 columns. While **Figure 4** below shows only 14 columns that have the largest value in importance for classifier purposes.

In the sample datasets, there are various scenarios of the data record. However, in this research, only the scenarios that are best suited to our purpose were used. The

labelz	x_acc	y_acc	z_acc	x_gyr	y_grc	z_grc	Azimuth	Pitch	Roll
1	3	-158	198	36	-16	-3	-41	-647	820
1	0	-155	197	33	-20	-4	-41	-653	815
1	2	-159	193	27	-24	-2	-44	-655	806
1	1	-154	192	21	-29	-3	-43	-654	802
1	0	-155	193	21	-32	0	-41	-658	799
1	0	-156	192	25	-37	-1	-36	-651	780
1	0	-155	189	26	-36	0	-35	-648	774
1	0	-154	190	26	-33	-1	-35	-644	774
1	3	-154	185	28	-27	0	-34	-645	772
1	5	-157	189	30	-19	0	-35	-647	772
1	0	-151	190	33	-8	1	-35	-641	778
0	10	-249	2	43	7	-18	12	-1021	25
0	12	-248	2	43	6	-18	16	-1018	21
0	12	-249	4	44	5	-17	19	-1020	20
0	4	-245	1	45	4	-20	13	-1018	29
0	5	-250	4	44	3	-20	18	-1018	23
0	12	-246	3	45	3	-19	18	-1022	18
0	10	-248	0	45	3	-20	13	-1015	20
0	8	-246	0	45	3	-21	17	-1016	20
0	7	-244	2	43	2	-21	14	-1007	18
0	12	-246	1	40	0	-20	10	-1012	18

Figure 2.
Sample contents of the Sisfall dataset.

Labelz	acc_x	acc_y	acc_z	gyro_x	gyro_y	gyro_z	azimuth	pitch	roll
0.0	4.400358792	9.045477364	-0.978305157	-0.112704635	0.12461651	0.04153884	8.672123	-77.34854	9.950522
0.0	4.39834008	9.051374883	-1.016111171	-0.12217305	0.13561209	0.043066	8.690639	-77.39494	10.205124
0.0	4.432260689	9.068335099	-1.028831443	-0.12797627	0.14813483	0.04184427	8.705528	-77.41798	10.457262
0.0	4.468305331	9.086357326	-1.042348231	-0.14355333	0.15057829	0.031154128	8.720216	-77.42717	10.702237
0.0	4.482851485	9.092582201	-1.048065114	-0.14966199	0.15454891	0.043066	8.728849	-77.41297	10.948666
0.0	4.494711692	9.097326383	-1.052809197	-0.1572978	0.1554652	0.039095376	8.73727	-77.38568	11.183537
0.0	4.506732399	9.102134767	-1.05761748	-0.17287487	0.15851954	0.048258353	8.745609	-77.3305	11.428347
0.0	4.51868124	9.106914403	-1.062397016	-0.1863139	0.17928895	0.051618114	8.747314	-77.26231	11.665751
0.0	4.511702814	9.10326799	-1.050200793	-0.19761491	0.18142697	0.055588737	8.751848	-77.176674	11.902264
0.0	4.502083947	9.098458481	-1.035772518	-0.22602014	0.17501289	0.057421334	8.751475	-77.07572	12.134822
1.0	4.507531878	9.10225466	-0.965677928	-0.23151793	0.18661933	0.05955936	8.743968	-76.95828	12.368984
1.0	4.51456682	9.106944703	-0.890638756	-0.24709499	0.19517145	0.06719518	8.737816	-76.82661	12.60147
1.0	4.52177884	9.1117528	-0.813710762	-0.27305678	0.22082779	0.070554934	8.716235	-76.68724	12.826323
1.0	4.528767034	9.116411677	-0.73917024	-0.29718593	0.21930063	0.07422013	8.699064	-76.53871	13.050086
1.0	4.521711785	9.149590858	-0.683189607	-0.3182608	0.23976462	0.077579886	8.667083	-76.38281	13.270337
1.0	4.512411733	9.186791069	-0.632039017	-0.34422258	0.25228736	0.09285152	8.636908	-76.21684	13.491641
1.0	4.506780825	9.21320947	-0.575751265	-0.36651915	0.28191432	0.106595986	8.588048	-76.03846	13.70991
1.0	4.502159794	9.236314141	-0.52030001	-0.39461896	0.29993483	0.10781772	8.538342	-75.856964	13.925496
1.0	4.527114929	9.267634045	-0.466688604	-0.42485678	0.33536503	0.11789699	8.473187	-75.6693	14.139905
1.0	4.560540076	9.301059342	-0.414163422	-0.45356745	0.35216382	0.124311075	8.407973	-75.481895	14.350269
1.0	4.58084015	9.290508215	-0.327819909	-0.46792278	0.3735441	0.1301143	8.325914	-75.293465	14.556428
1.0	4.59725424	9.267059514	-0.231680169	-0.49357912	0.4025602	0.13958271	8.236505	-75.102325	14.761983

Figure 3.
Sample contents of the MobiAct dataset.

1 - fBodyBodyGyroJerkMag-entropy()	2 - fBodyBodyGyroJerkMag-maxInds	3 - fBodyBodyGyroJerkMag-meanFreq()	4 - fBodyBodyGyroJerkMag-skewness()	5 - fBodyBodyGyroJerkMag-kurtosis()	6 - angle(tBodyAccMean,gravity)	7 - angle(tBodyAccJerkMean),gravityMean)	8 - angle(tBodyGyroMean,gravityMean)	9 - angle(tBodyGyroJerkMean,gravityMean)	10 - angle(X,gravityMean)	11 - angle(Y,gravityMean)	12 - angle(Z,gravityMean)	13 - ActivityName	14 - Records Per Activity
0.47737	-0.96825	0.00207	-0.29942	-0.69117	-0.29479	-0.58219	0.04613	-0.11267	-0.76834	0.25671	0.04047	WALKING_DOWNSTAIRS	986
0.41023	-0.90476	-0.00661	-0.33147	-0.68595	0.1988	-0.51506	-0.96081	0.79859	-0.73018	0.27574	0.07579	WALKING_UPSTAIRS	1073
0.53831	-0.90476	-0.15164	-0.31214	-0.67395	-0.16088	0.56474	0.91247	-0.41284	-0.77371	0.25243	0.0441	WALKING	1226
-0.46849	-1.0	-0.27489	0.38391	-0.02176	-0.06858	-0.04311	-0.33225	0.48751	0.43248	-0.55335	-0.43288	LAYING	1286
-0.79589	-1.0	-0.25761	0.15619	-0.24178	0.01353	0.04335	0.02149	0.04669	-0.66708	0.05422	-0.21887	SITTING	1374
-0.87131	-1.0	-0.07432	-0.29868	-0.7103	-0.11275	0.0304	-0.46476	-0.01845	-0.84125	0.17994	-0.05863	STANDING	1407

Figure 4.
Sample contents of the Ucihar dataset.

human activities include that were studied include, Standing, Sitting, Laying, walking fast, walking slow, walking downstairs, and walking upstairs [24]. We divided these activities into two groups per dataset. Group 1 contains ADLs and group 0 contains falls. There are four types of fall activities and six types of ADLs. For data preprocessing, an operation is performed to divide records into two groups. In the fall detection procedure, MobiAct and Sisfall datasets will groups ADLs as zeros and all the several types of falls as ones. In the activity recognition procedure, the data in the Ucihar dataset's group consisting of Laying activity (sleeping) is labeled as 1, and all other activities are grouped under label 0. By creating these two groups we can thus use a simpler classifier, binary classification, instead of multiple classifications.

3.4 Algorithms selection from common algorithms

Several algorithms exist for anomaly detection systems. For the execution of detection falls, we shall utilize eight different machine learning algorithms [25]. These algorithms would be compared, and the best should be used in the application of the smart house. Eight algorithms are selected for this experiment as follows. Logistic regression, Linear discriminate analysis (LAD), k-nearest neighbors (k-NN), decision tree classifier, Gaussian naive Bayes, Support Vector Machine (SVM), Random Forest, and xgboost algorithms. The best performing among these algorithms is to be utilized. The algorithm's performance would have to be weighed by the following parameters.

a. Optimization—eliminate the worst performing algorithms

b. Completeness—eliminate some other solution where a result is returned

c. Accuracy and precision—the degree of accuracy attained and required

d. Execution time—period of performance the classification

e. Resource consumption—memory and processor usage

The combination of human activity detection and the fall detection algorithm is detected based on these above four factors utilized.

3.5 Design of the floor plan of the smart house

Figure 5 below which shows the simulated floor plan of the smart house. The sample house has the following floor plan. A1 and A2 are bedrooms for sleeping. B1 and B2 are corridors. C is the toilet; D is the Bathroom and E is the kitchen for cooking.

3.6 Remainder alarm for medicine taking routine and camera/pressure mat

There would be a remainder device. This becomes the third Component of the system. The first is fall detection, the second is activity classification and the third is alarm detection and camera. Given the apartment above, we must use the fall data and the human activity recognition. We should make these assumptions.

1. Non-serious fall is considered a fall. No alarm but warning recorded

2. Sleeping occurs in room A1 or A2. Send alarm if sleeping anywhere else.

3. Laying outside of rooms A1 and A2 should trigger a warning

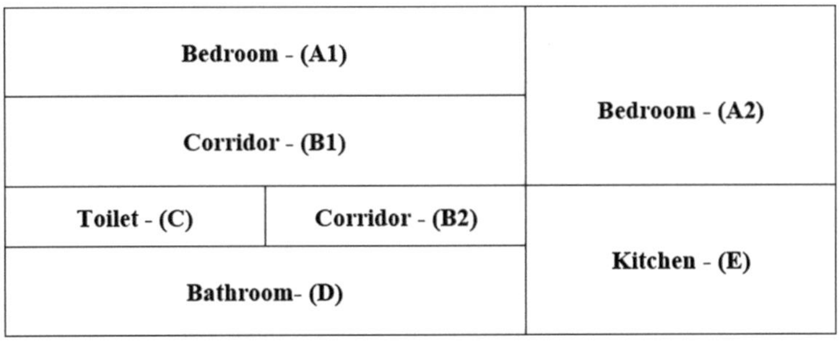

Figure 5.
An illustration of the sample smart house floor plan design.

4. Activity should not switch abruptly. For example, sleep to walk to walk

5. Being in the toilet for a lengthy period indicates a problem, hence alarm is evoked,

6. A medical dispensary is kept in the room (E). A pressure sensor sends a photo when a pill is removed from the dispensary. After 5 minutes after the scheduled time for taking the pill passes, then an alert is sent out to the caregiver.

These six points are used as the heuristic when identified. Once it is identified then an alarm or warning is evoked. We need to conduct two tests before we send an alarm as a way of avoiding false alarms. Identify the activity and then identify the room and the applicable algorithm's location also establish that Laying is not in an inappropriate room. Furthermore, a delay in the bathroom should trigger a warning. The second algorithm verifies that a fall is not a Laying, and a Laying is not a fall. Once this is established only then can an alarm to send. This could reduce false alarms and increase confidence levels unlike having one algorithm.

3.7 Description of the three algorithms

Three algorithms are derived to execute the above procedures. Below is the description of the three algorithms pseudocode.

Algorithm 1: Update datasets, test, and use them in future training.

01: Retrieve sample records from the datasets #1, #2, and #3.
02: train model using standard algorithms and records of the subject person.
03: If the time elapses sent collected data for analysis by the selected algorithms.
04: end if.
05: if the algorithm accuracy is top two use the best and then discard the rest.
06: end if.
07: process and identify the chances of ADL.
08: process and identify the chances of a FALL.
09: if ADL is sleeping but the room is not sleeping quarters send an alert signal.
10: endif.
11: if a Fall is detected sent an alert signal.
12. else save the data and then move into the waiting stage.
13: endif.
14. Repeat the strategy starting from point 01.

Algorithm 2: Sleeping area locator for logical heuristic missed fall prediction.

01: Retrieve sample records from the datasets #1, #2, and #3.
02: train model using standard algorithms and records of the subject person.
03: if the HAR is laying or napping find out which room is activity occurring.
04. If the room is not appropriate for Sleeping quarters, send a warning alarm.
05. If sleep is in the sleeping quarter's location, then move to the waiting stage.
06: end if.
07. Repeat the strategy starting from point 01.

Algorithm 3: Medical remainder algorithm, to predict skipping of medicine routine.

01: Retrieve from schedule records on times required for taking medication.
02: At each required time check the weight of the medicine pressure.
03: If the weight has been adjusted then its confirmed medicine has been taken.
04: If the weight is still the same then send a warning signal to inform caretaker.
05. else save the data and get a new data sample.
06: endif.
07. Repeat the strategy starting from point 01.

Based on the above algorithms, the alarm is triggered as a response. These responses will differentiate possible similar activities (such as laying and falling) before evoking the alarm. As shown in **Table 3**, when the results of the algorithm are as provided in Answer (I) then the Alarm is evoked. If the answers are as in Answer (II) then a warning is logged in a database. Three warnings in a sequence also trigger an alarm.

3.8 Updating training dataset

The data collected from the senior citizen's sensor is originally not labeled. When a detection process is completed the sensor data would then be assigned a label. Once labeled, then the system would save this information with its given label. After the label is authenticated, this record is then moved to the created dataset for extension of the original dataset. At this point, the system would save the labeled data into a new dataset which is the original dataset plus the new record. The record can then be used in training sessions. This new dataset would have an extra record that more closely represent the person involved. In this case, the training would reflect the subject senior citizen. Below is **Figure 6** showing the systems' flow chart.

In Article [3] similar research is presented. The authors make a comparison of three datasets and look at the performances of different algorithms. In this work, we have compared results from two datasets for the human activity detection algorithms. The results would be fused to reduce the probability of false positive or false negative.

4. Results

Eight algorithms had their performance studies as indicated in **Figure 7** below. Of these eight, some would be discarded in preference for the best-performing algorithm.

Algorithms	Tests before triggering an alarm	Answer(I)	Answer (II)
1	Falling has been detected or not?	Yes	No
2	Laying has been detected or not?	No	Yes
3	Laying occurring in an appropriate room?	No	Yes
Result	Send alarm to Caregiver	Yes	No

Table 3.
Question to answer before invoking a particular algorithm.

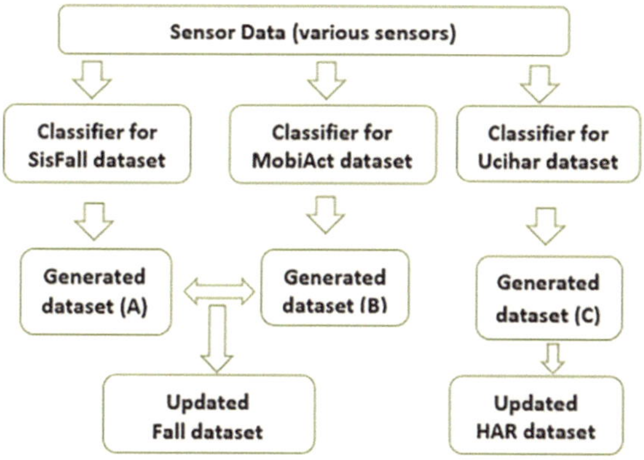

Figure 6.
Flow chart of creation of custom datasets.

	ALGORITHM USED	SISFALL	MOBIACT	UCIHAR
10001	LogRegression Accuracy	0.7082242012000001	0.7993713986	0.7959183673000001
10002	LDAnalysis Accuracy	0.7380827658	0.7962283918	0.7972789116
10003	KNNeighbour Accuracy	0.992142483	0.9973808276999999	0.9156462584999999
10004	DTClassifier Accuracy	0.9785227868	0.9905709796	0.8979591837
10005	GNBayes Accuracy	0.760607648	0.9209009953	0.6639455782
10006	SVMachine Accuracy	0.9559979047	0.8349921425	0.8013605442
10007	RandomForest Accuracy	0.9832372970000001	0.9979046621	0.9346938776
10008	XGBoost Accuracy	0.9874279728	0.9979046621	0.9414965986

Figure 7.
Comparison of algorithm performance per dataset.

During the experiment, algorithm ranking was established. The lower-ranked as less effective algorithms are to be eliminated. This reduced number of options increases the efficiency as unfavorable options are not computed. The extra computation would be the worst of resources. The selecting of the best option thus avoids unnecessary use of computing power in analyzing some irrelevant options. The less likely algorithm options are removed immediately when identified. Below is **Figure 7** which is a snapshot of the accuracies of various investigated methods.

In **Figure 7**, Algorithm KNN is more accurate for the Sisfall dataset, and XGBoost is more accurate for both MobiAct and Ucihar datasets. A selection of three of the graphs of the accuracy with the best performance is presented below. The next two graphs are for fall classification and the second is for human activity classification.

The detection accuracy for the Sisfall dataset is indicated as 98.08% for the training session and 97.92% for the testing session as shown below in **Figure 8**. This is the best accuracy from the list of classifiers in the experiment.

The detection accuracy for MobiAct dataset accuracy is indicated in **Figure 9** at 99.23% for the training session and 98.84% for the testing session. This is the best accuracy from the list of classifiers in the experiment.

The detection accuracy for Ucihar dataset accuracy is indicated at 96.85% for the training session and 95.21% for the testing session as shown below in **Figure 10**. The Ucihar accuracy at 96.85% is the worst accuracy of the usages of XGBoost classifiers.

The Sisfall test dataset has 14,783 ADLs cases and 3775 fall cases; **Figure 11** shows the confusion matrix for the Sisfall test dataset during the training session.

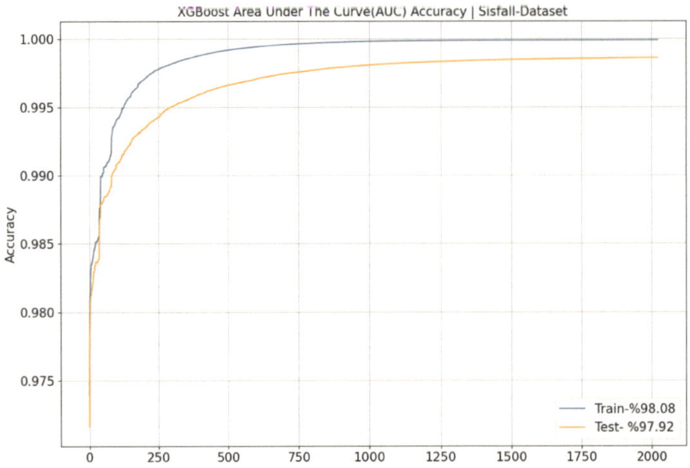

Figure 8.
Results for accuracy simulation using the Sisfall dataset.

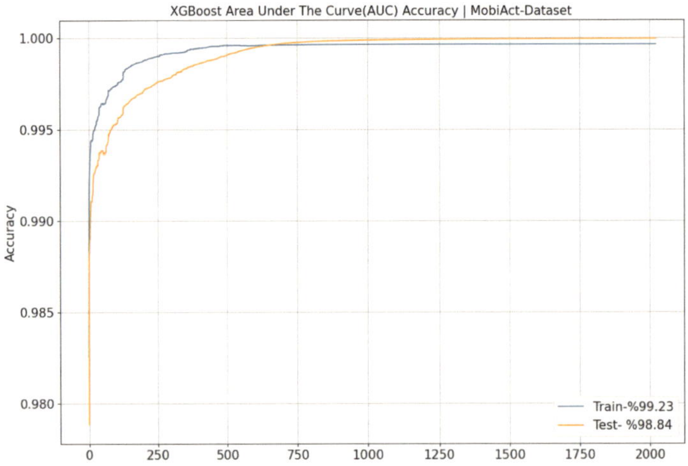

Figure 9.
Results for accuracy simulation using MobiAct dataset.

The MobiAct test dataset has 16,562 ADLs cases and 3984 fall cases; **Figure 12** shows the confusion matrix for the MobiAct test dataset.

The Ucihar test dataset has 1963 ADLs cases and 464 Laying cases; **Figure 13** shows the confusion matrix for the Ucihar test dataset.

From the confusion matrix, we can extract the accuracy, sensitivity, and specificity of our classifier. The higher each of these parameters the better the performance of the classifier. However, accuracy must be considered in conjunction with specificity and sensitivity. A classifier must have a high sensitivity and specificity, to be defined as having superior performance. As seen in **Table 4** below, both sensitivity and specificity are above 78% which is high performing case. This shows that the one selected option from the eight models performs quite well and can be used to develop the proposed system. As indicated in **Figure 10**, the MobiAct dataset accuracy is recorded at 98% for the training and at 99% for the testing which is the best accuracy of our

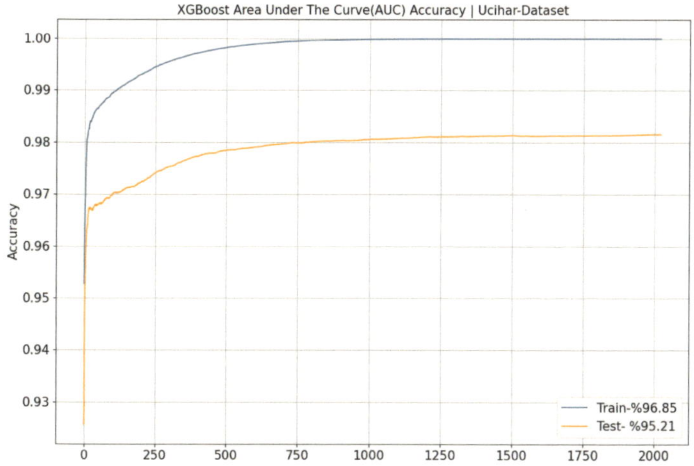

Figure 10.
Results for the accuracy simulation using the Ucihar dataset.

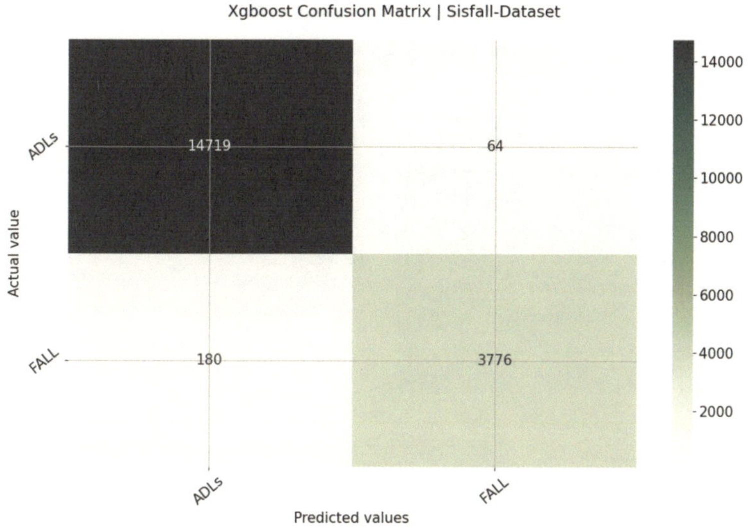

Figure 11.
Confusion matrix for XGBoost classifier on Sisfall dataset.

possible classifiers. Below is **Table 4** which indicates the performance of the most effective classier(XGboost) in the experiment.

The high accuracy indicates that the models were efficient and can be used in detections. However, when executing these algorithms speed and accuracy are a factor in optimization. If given more time an algorithm can perform better. However, this has to be considered with efficiency on time when an anomaly is reported. If it takes too much time to compute a high accuracy decision, it could be that by the time the decision is taken it's too late for the damage already done. Below is **Figure 14** which indicates the time it took for each algorithm to complete a single task.

Moreover, these eight machine learning methods were compared with deep learning. Deep learning has multiple models chained together to enhance performance. However, the computation costs were more than for the machine learning method.

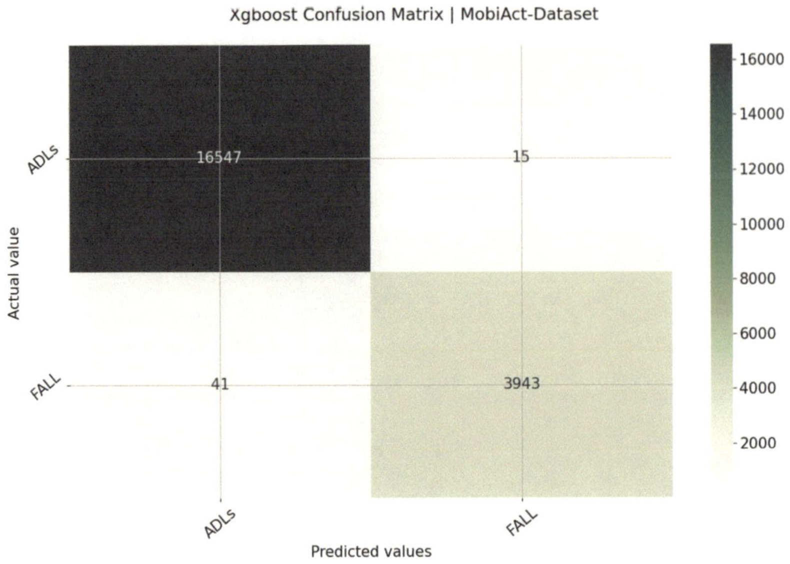

Figure 12.
Confusion matrix for XGBoost classifiers on MobiAct dataset.

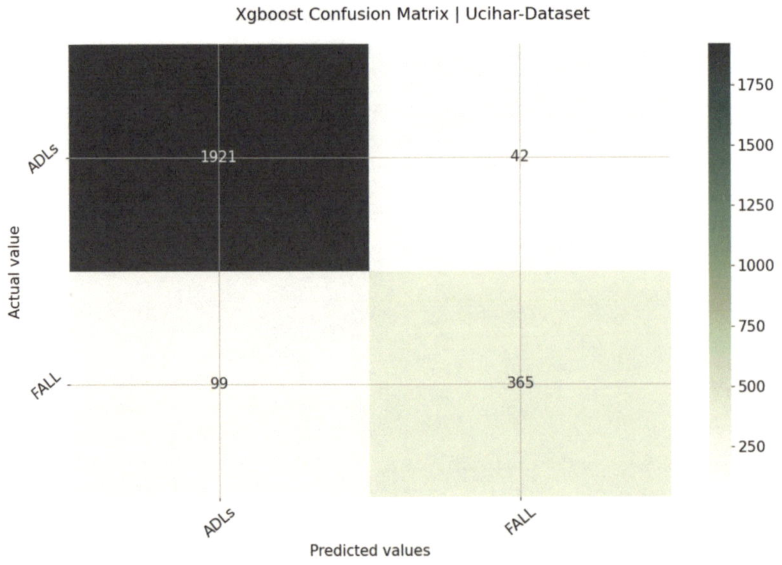

Figure 13.
Confusion matrix for XGBoost classifier for Ucihar dataset.

The accuracy deep learning method on the Sisfall dataset Sisfall selection dataset accuracy is recorded at 96.84% for the training session and at 93.55% for the validation session as indicated below in **Figure 15**.

Deep learning accuracy on the MobiAct dataset selection dataset accuracy is recorded at 96.97% for the training session and at 100.0% for the validation session as indicated below in **Figure 16**.

Dataset Extracts	True / False	False / True	Class Total	Final Total	Accuracy	Sensitivity	Specificity
ADL(Sisfall)	**14,719**	**64**	14,783	18,739	98.69%	95.42%	99.57%
Fall (Sisfall)	**181**	**3775**	3956				
ADL(MobiAct)	**16,547**	**15**	16,562	20,546	99.73%	98.97%	99.91%
Fall (MobiAct)	**41**	**3943**	3984				
ADL(Ucihar)	**1921**	**42**	1963	2427	94.19%	78.66%	97.86%
Laying(ucih..)	**99**	**365**	464	2427			

Table 4.
Indicators of classifier efficiency.

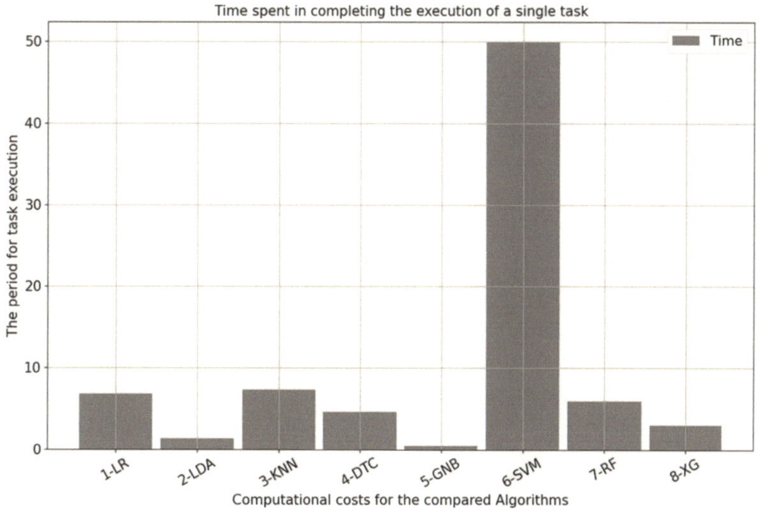

Figure 14.
Recorded time consumed per tested algorithm.

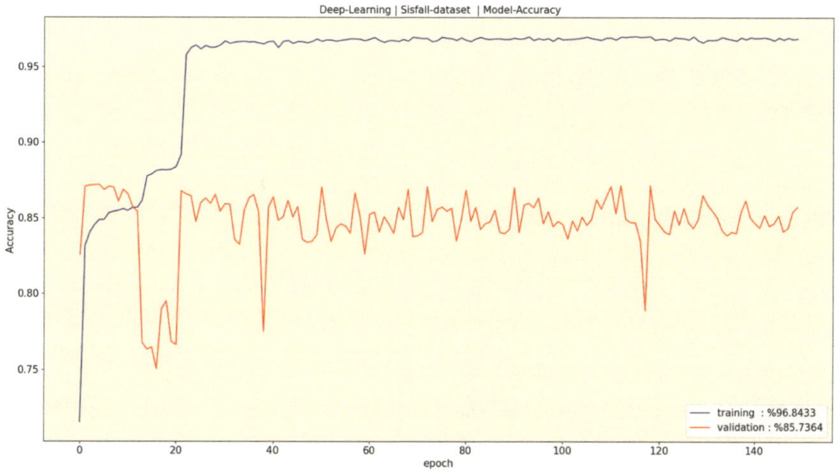

Figure 15.
Deep learning accuracy on the Sisfall dataset.

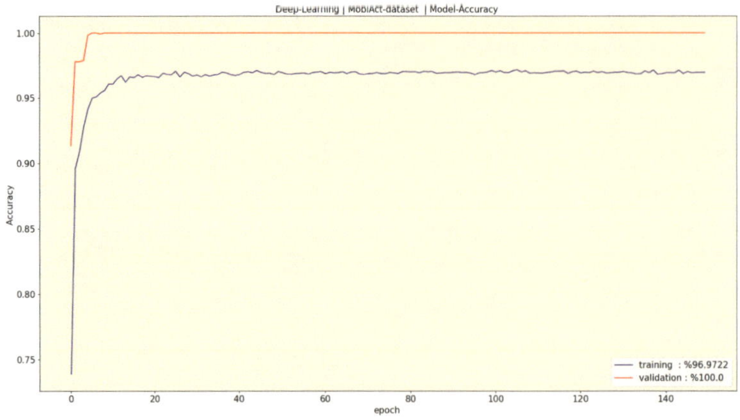

Figure 16.
Deep learning accuracy on the MobiAct dataset.

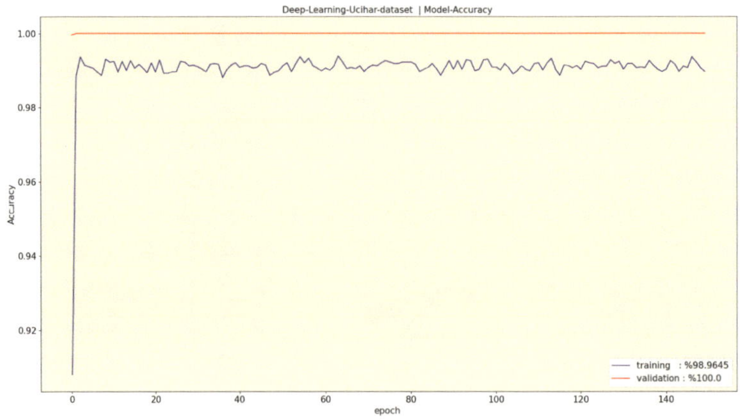

Figure 17.
Deep learning accuracy on the Ucihar dataset.

Deep learning accuracy on the Ucihar dataset Sisfall selection dataset accuracy is recorded at 98.96% for the training session and at 100.0% for validation as indicated below in **Figure 17**.

5. Discussions and future works

The results show that the fall detection and activity recognition algorithms are competitive. The fall detection was at 96% minimum accuracy, which is above average for the cited other researchers' works which had an average rate of 89% as shown in **Table 1**. We have used cross verification with other dataset records. This is with the hope that the subject senior citizens' data would have to be recorded and then observed to make the correct adjustments. The activity recognition was the most accurate. It performed at 98.96% for training and 100.0% for the validation session. The general accuracy for the xgboost algorithm was above 96% on each of the three datasets. However, the deep learning method was the most accurate. After a successful classification, we can add the new record and its classification label to the dataset. This thus extends and improves the dataset. We could also include one heuristic to improve accuracy and reduce computational costs. For instance, we should be able to tell that sleeping must have zero chance of occurring in the toilet, hence not consider it a possibility when computing. This allows resources to be concentrated on viable options when performing classification. Eliminating such computation decision cost can fine-turn the system, as it allows only options with high likelihood. The low likelihood options are removed from the allocation of computation resources. This allocation of energy to the option improves efficiency. In the case that one class is larger than another the dataset is unbalanced. In usual cases this system works on unbalanced datasets, hence it was important to have good characteristics in its sensitivity and specificity.

5.1 Edge processing and security

In this work, the data sensing was performed using non-protruding methods which are a mobile phone and a smartwatch data sensor. However, this method is replaceable in the structure of the system. The data can also be collected by using a camera. A labeled dataset is then used to label extracted images. Mobile collects data and then this data is not processed on the phone but sent to some processing point due to limitations of the processing capability of the phone [26]. The training and testing set labels the images. However, in our project, we use an accelerometer and gyroscope as input sensors. The rest of the system would be the same. A systems camera sensor is an advantage in that it would be easier to label.

5.2 Replacement of training and testing dataset

When data is collected from the subject because seniors change rapidly for deteriorate help, the trend of that senior would change. Hence the dataset must be properly monitored and adjusted to match the rate of changes in the trend of the senior. Otherwise maintaining the same training data for an extended period would result in an obsolete detection system [20]. Aging can change the gait pattern of an individual hence the importance to update the dataset constantly. Having been limited by the current covid-19 situation we are having difficulty arranging our data collection activities. However, we believe the public datasets, SisFall, MobiAct, and Ucihar have provided good insights into the record we could have managed to collect. We believe these datasets have an unobstructed view of the results we could have obtained. After collecting sensor data and labeling it, the dataset component from Sisfall, MobiAct, and Ucihar databases would be gently removed and replaced with these records. This is a continuous process until the dataset remains pure containing the new record of the senior citizens without the legacy dataset. Erroneous labeled data would continuously be removed to have a robust and current training dataset.

5.3 The medicine routine remainder service

The medicine remainder is a time-based scheduler. The schedule must be executed correctly and if not, then an alarm is sent. The alarm is triggered based on the failure of sending confirmation of executing the medicine schedule by the senior. The senior must send a confirmation once prompted to do so. However, there is a risk that the subject might be able to falsely confirm they took the medicine when in fact they did not take it. This system currently cannot help when the senior is specifically not providing the correct state of medicine routine. It is meant to help in cases where participants are willing to take the medicine. An alert will be sent to caregivers allowing them to prompt the senior on the state of their medicine routine.

6. Conclusion

We commend this system can evaluate and monitor the situation and status of the senior citizen in their apartment. This system has been designed using three main algorithms. First algorithm tests if a fall has occurred or not. Eight different common algorithms are evaluated to see which one is most effective. If a fall is detected, then the caregiver is alerted. The second algorithm is to identify what activity is the senior citizen doing. This algorithm has the purpose of detecting the location where activity is occurring. If an activity such as sleeping or laying is taking place in the wrong location an alert is evoked. The last algorithm is detecting whether the medicine routine prescribed is been executed or not. The system utilizes a dispenser that can record if a pill was extracted from the pill dispenser or not. This is checked at a specified period as scheduled. When there is no change in the medical container after the expected pill dispensation time elapsed, then an alert warning message is sent to caregivers. The medical remainder also enables caregivers to prompt subject seniors to implement the medicine routine, when medicine taking has been skipped.

These three algorithms are the primary functions of this smart house design. The system utilizes the executed algorithms to effect detection. The system must use few resources and utilize the improved performance algorithm. The design would be updated with specific data when training for a specific client. The sample, data is also to be replaced with the change in the pattern of the senior citizen trends. The final efficiency in the algorithms improves from 96–98% in the training session and the validation is at 85–100% for the testing session. Since the customized records are generated from the citizen, the training and validation are guaranteed to improve at every iteration. With a sensitivity minimum of 78.66% and a specificity minimum of 97.86%, the model is performing well as the dataset used is not balanced.

This prototype would allow using the most effective classifier and dynamically determine the most effective classifier. Dynamic evaluation of algorithm efficiency should be integrated, as the accuracy would not always be the best since training data is updated periodically. Using the location variable may also reduce the computing resources needed if integrated into the classifier algorithm. Without a logic heuristic, the computation process would require more resources.

Acknowledgements

This work was supported by IGA/CebiaTech/2021/001, a research project of the Faculty of Applied Informatics, Tomas Bata University in Zlín.

DOI: http://dx.doi.org/10.5772/intechopen.103953

Author details

Linos Nchena* and Martin Tomášek
Tomas Bata University, Zlín, Czech Republic

*Address all correspondence to: nchena@utb.cz

References

[1] World Health Organization. Active Aging: A Policy Framework. No. WHO/NMH/NPH/02.8. Madrid Spain: World Health Organization; 2002

[2] Sander M, Oxlund B, Jespersen A, Krasnik A, Mortensen EL, Westendorp RGJ, et al. The challenges of human population ageing. Age and Ageing. 2015;**44**(2):185-187

[3] Kaluža B, Mirchevska V, Dovgan E, Luštrek M, Gams M. An agent-based approach to care in independent living. In: International Joint Conference on Ambient Intelligence. Berlin, Heidelberg: Springer; 2010. pp. 177-186

[4] Islam MM, Tayan O, Islam MR, Islam MS, Nooruddin S, Kabir MN, et al. Deep learning based systems developed for fall detection: A review. IEEE Access. 2020;**8**:166117-166137

[5] Reyes-Ortiz J-L, Oneto L, Sama A, Parra X, Anguita D. Transition-aware human activity recognition using smartphones. Neurocomputing. 2016; **171**:754-767

[6] He J, Zhang Z, Wang X, Yang S. A low power fall sensing technology based on FD-CNN. IEEE Sensors Journal. 2019;**19**(13):5110-5118

[7] Xu T, Se H, Liu J. A two-step fall detection algorithm combining threshold-based method and convolutional neural network. Metrology and Measurement Systems. 2021;**28**(1):23-40

[8] Usmani S, Saboor A, Haris M, Khan MA, Park H. Latest research trends in fall detection and prevention using machine learning: A systematic review. Sensors. 2021;**21**(15):5134

[9] Zhu C, Sheng W, Liu M. Wearable sensor-based behavioral anomaly detection in smart assisted living systems. IEEE Transactions on Automation Science and Engineering. 2015;**12**(4):1225-1234

[10] Paudel R, Eberle W, Holder LB. Anomaly detection of elderly patient activities in smart homes using a graph-based approach. In: Proceedings of the 2018 International Conference on Data Science. United States: CSREA Press; 2018. pp. 163-169. ISBN: 1-60132-481-2

[11] Hussain F, Umair M, Ehatisham-Ul-Haq M, Pires I, Valente T, Garcia N, et al, editors. An efficient machine learning-based elderly fall detection algorithm. In: SENSORDEVICES 2018, the Ninth International Conference on Sensor Device Technologies and Applications, Venice, Italy, 16–20 September 2018. United States of America: Xpert Publishing Services; 2018

[12] Vavoulas G, Pediaditis M, Chatzaki C, Spanakis EG, Tsiknakis M. The mobifall dataset: Fall detection and classification with a smartphone. International Journal of Monitoring and Surveillance Technologies Research (IJMSTR). 2014;**2**(1):44-56

[13] Özdemir AT, Barshan B. Detecting falls with wearable sensors using machine learning techniques. Sensors. 2014;**14**(6):10691-10708

[14] Bouchabou D, Nguyen SM, Lohr C, LeDuc B, Kanellos I. A survey of human activity recognition in smart homes based on IoT sensors algorithms: Taxonomies, challenges, and opportunities with deep learning. Sensors. 2021;**21**(18):6037

[15] Wisesa IWW, Genggam Mahardika. Fall detection algorithm based on accelerometer and gyroscope sensor data using recurrent neural networks. In IOP Conference Series: Earth and Environmental Science. Vol. 258, No. 1. United Kingdom: IOP Publishing; 2019. p. 012035

[16] Jin M, Zou H, Weekly K, Jia R, Bayen AM, Spanos CJ. Environmental sensing by wearable device for indoor activity and location estimation. In: IECON 2014-40th Annual Conference of the IEEE Industrial Electronics Society. United States of America: IEEE; 2014. pp. 5369-5375

[17] Luo Z, Hsieh J-T, Balachandar N, Yeung S, Pusiol G, Luxenberg J, et al. Computer vision-based descriptive analytics of seniors' daily activities for long-term health monitoring. Machine Learning for Healthcare (MLHC). 2018;**2**:1

[18] Venkatraman S, Overmars A, Thong M. Smart home automation—Use cases of a secure and integrated voice-control system. Systems. 2021;**9**(4):77

[19] Mahfuz S, Isah H, Zulkernine F, Nicholls P. Detecting irregular patterns in IoT streaming data for fall detection. In: 2018 IEEE 9th Annual Information Technology, Electronics and Mobile Communication Conference (IEMCON). United States of America: IEEE; 2018. pp. 588-594

[20] Delahoz YS, Labrador MA. Survey on fall detection and fall prevention using wearable and external sensors. Sensors. 2014;**14**(10):19806-19842

[21] Sucerquia A, López JD, Vargas-Bonilla JF. SisFall: A fall and movement dataset. Sensors. 2017;**17**(1):198

[22] Vavoulas G, Chatzaki C, Malliotakis T, Pediaditis M, Tsiknakis M. The MobiAct dataset: Recognition of activities of daily living using smartphones. In: International Conference on Information and Communication Technologies for Ageing Well and e-Health. Vol. 2. Portugal: SCITEPRESS; 2016. pp. 143-151

[23] Anguita D, Ghio A, Oneto L, Parra X, Reyes-Ortiz JL. Energy efficient smartphone-based activity recognition using fixed-point arithmetic. Journal of Universal Computer Science. 2013;**19**(9): 1295-1314

[24] Liu L, Hou Y, He J, Lungu J, Dong R. An energy-efficient fall detection method based on FD-DNN for elderly people. Sensors. 2020;**20**(15):4192

[25] Thakur N, Han CY. A study of fall detection in assisted living: Identifying and improving the optimal machine learning method. Journal of Sensor and Actuator Networks. 2021;**10**(3):39

[26] Pan D, Liu H, Dongming Q, Zhang Z. CNN-based fall detection strategy with edge computing scheduling in smart cities. Electronics. 2020;**9**(11): 1780

Chapter 6

FAIME: A Framework for AI-Assisted Musical Devices

Miguel Civit, Luis Muñoz-Saavedra, Francisco Cuadrado, Charles Tijus and María José Escalona

Abstract

In this paper, we present a novel framework for the study and design of AI-assisted musical devices (AIMEs). Initially, we present taxonomy of these devices and illustrate it with a set of scenarios and personas. Later, we propose a generic architecture for the implementation of AIMEs and present some examples from the scenarios. We show that the proposed framework and architecture are a valid tool for the study of intelligent musical devices.

Keywords: artificial intelligence, musical devices, internet of musical things

1. Introduction

Advances in technology and computer science have greatly enhanced the possibility of designing, developing, and deploying intelligent musical devices. A typical well-studied subset of these intelligent devices are IoMusTs (Internet of Musical Things). According to [1], an IoMusT is a "computing device capable of sensing and exchanging data to serve a musical purpose." An IoMusT does not need to be able to produce, select, or modify music, but it can be any device that is "music aware" in the sense that its behavior is directly related to music. As an example, PixMob devices [2] have been widely used in musical performances. These devices that can be either worn (smartband), thrown (balls), or attached to the audience seats and are able to produce light patterns synchronized with life performances.

Not all intelligent musical devices are IoMusts. We can design intelligent devices where the intelligence is embedded in the device, and thus, we may say that we gave an Intelligent Musical dEvice (IME) but not as a part of the Internet of Musical Things. In [3], the evolution of the design of intelligent musical instruments is studied. In most cases, these instruments use artificial intelligence as a tool for user interaction without requiring any connection to public networks or cloud-based services. It is important to consider that machine learning ML, in most of these cases, cannot be considered as an independent agent but mainly as one of the possible alternatives for designing layers of a complete system. These types of devices can also be considered as cyber-physical systems, as they clearly require intelligent software systems and dedicated hardware.

In this work, we will create a framework that covers all, or at least a very wide part, of intelligent musical devices and helps design, understand, and study them.

The rest of the paper is divided as follows: First, in the Materials and Methods section, the taxonomy is and published works are detailed, as well as the analysis methodology used to test the different systems. The results obtained for the different systems are then detailed and explained in the Results and Discussion sections. Finally, conclusions are presented.

2. Materials and methods

Artificial-intelligence-assisted musical devices come in a wide variety of forms and potentially have a very wide spectrum of uses. In order to create a framework that will cover most of these possibilities, we will start by introducing taxonomy of the different usages of the said devices. It should be clear that it is possible for a device to fall into several categories. As an example, most musical instruments could also be considered educational aids, some of them being used predominantly for this purpose. The monochord was used through the Middle Ages for educational and scientific purposes [4], and similarly, we can design intelligent instruments that, although being able to be used for performing, are meant with an educational intent.

2.1 Taxonomy

We propose a classification for AI-assisted musical devices (AIMEs). It is clear that this is not the only possible taxonomy, but it is complete, easy to apply, and useful. The classification is shown in **Table 1**.

In a first level, we divide our AIMEs into:

- Devices that are played by musicians: Musical instruments.
- Devices designed to modify music: Music Processors.
- Devices that compose music: Music Generators.
- Devices that select music: Music Recommenders.
- Devices that send to the user or the environment information extracted from the music: Feedback systems.
- Devices designed to be used in an educational process: educational devices.

A real device may be included in several categories. As an example, a device could generate a set of music scores and then recommend some of them to a student. In this way, this device could be considered as a generator, a recommender, and an educational system.

This main AIME division can then be divided into subcategories. As an example, a Music Generator can either be instrumental, vocal, or combined. An instrumental music generator usually produces music in symbolic format. The most common symbolic format is the Musical Instrument Digital Interface (MIDI), which contains information that indicates the pitch, start time, stop time, and other properties of each individual note, rather than the resulting sound. Combined and voice generators have to use a raw audio format and are much more difficult to implement, although their quality has improved significantly in the present decade [5].

1. Musical instruments	a. AI assisted instruments b. Augmented instruments
2. Music processors	a. Instrumental modifiers b. Voice modifiers c. General sound processors
3. Music generators	a. Instrumental b. Voice c. Combined
4. Music recommendation devices	a. Ambient aware recommendation b. User aware recommendation c. Combined
5. Music-related feedback systems	a. Personal Feedback. b. Ambient Feedback. c. Combined
6. Educational Aids	a. Music Education b. General educational support c. Rehabilitation

Table 1.
AI-assisted musical device (AIME) taxonomy.

As a further example, recommendation devices can recommend music as a function of the environment or as a function of the user state. The environment-based recommendation is mostly used in social scenarios, e.g., if the system selects music for a shopping mall or an elevator. Personal Music recommendation devices are used mostly when recommending for a single user. As an example, we could estimate the user's emotional state from the data of the wearable device [6] and select the music accordingly. It is also possible to use the acquired data of an AIME personal recommender to try to modify some aspects of user behavior. An interesting possibility would be to train the user, through music, to reduce his or her stress level. In this way, the device could also be considered as part of the Internet of Behavior (IOB) [7].

2.2 Intelligent instrument scenarios

The area of intelligent musical instruments [8] includes an important subset of musical devices and has a wide range of applications that we will present in four example scenarios.

2.2.1 Able instrument scenario

Mike had an accident that led to a problem that prevents him from playing with his right hand. However, he would like to continue playing the bass in a small blues band. Mike thought he would not be able to play again as a bass player, as most instruments require significant ability with both hands. There are several alternatives to adapt the instrument to his physical capabilities [9], but finally he settled on a small robotic mechanism that can detect which string is he fretting with his left hand and pluck it. This device can hear what other members of the band are playing and dynamically adapt to the tempo and genre of the song by varying the rhythms and patterns it plays.

Although the results do not match his earlier performances, Mike is still able to play well enough and have fun with his friends' band.

2.2.2 Drum stroke scenario

Toby recently had a stroke that left him with reduced mobility in his right hand. In his rehabilitation clinic, they proposed that he should follow complementary music-supported therapy (MST) in which he controls a set of midi drums through his hand gestures [10], which are detected through electromyography signals (EMG). The drums can play almost autonomously at the beginning of therapy and allow control of an increased number of variables as Toby progresses in his recovery.

The rehabilitation device keeps track of Toby's progress and periodically sends reports to his therapist. When Toby goes to the clinic for an in-person session, the therapist will discuss his progress and adapt the MST accordingly.

2.2.3 Teach and play scenario

Mary wants to start playing the concertina and is following a well-known book and taking some lessons online. However, she does not like the sound that she is producing with the instrument currently and refuses to play it anywhere. A friend tells her about Inteltina, an intelligent didactical concertina that augments Mary's abilities and helps her produce a nice sound. The instrument assistance dynamically decreases as Mary's playing capabilities improve.

Although Mary plays reasonably well with Inteltina, her online teacher warns her that this type of instrument sometimes backfires as the student becomes lazy and her abilities stagnate [8].

2.2.4 TherAImin

Sara is a computer scientist who plays piano as a hobby. Recently, she has become fascinated by the discovery of the Theremin [11]. **Figure 1** shows an early implementation of Theremin. Being an AI specialist, she believes that the design can be clearly improved with the help of AI. Thus, she decides to become a "digital luthier" and to create a new instrument that is faithful to the original Theremin concept. The TherAImin keeps the pitch and volume antennas of the original instrument but includes an AI-based gesture recognizer to change the timbre of the instrument [12] according to hand gestures.

This scenario reflects the creation of new digital AI-supported musical instruments. Several interesting reflections on this topic can be found in [3].

This type of instrument is fun to build and play, but it can be difficult to create a community of users around them.

2.3 Audio processing scenarios

This area includes instrument processors, voice processors, and generic audio processors.

2.3.1 Boogie boogie scenario

Saul is a professional guitar player. He would love to have a Mesa Boogie Mark V amplifier, but the price is too high for him. Saul knows that there are emulations for

Figure 1.
Alexandra Stepanoff playing the theremin, 1930.

this amp for several Digital Audio Workstations (DAWs) including Cubase, which he regularly uses. However, Saul would like to have the emulation as a pedal he can easily carry. He has several friends who work in a small start-up company that designs embedded deep learning devices and learns from them that the boogie can be emulated by an AI system [13] that can be implemented using a Coral Edge TPU accelerator [14].

In a few months, Saul has tested the device and the company is starting to sell the BoogieBoogie Pedal.

2.3.2 DeepTuner Scenario

Sara is a singer who regularly uses a pitch-correction voice processor for her performances. Currently, she uses an AI enhanced version of Antares Auto-Tune [15] on an Avid Carbon Device. She is satisfied with the natural feeling, and virtually unnoticeable delay that this hardware/software implementation brings to her performances. Nevertheless, she would love a similar pitch correction implementation in a smaller and cheaper device [16].

2.3.3 DeepAFx scenario

Kyra is a production Engineer. Since she discovered the Deep-Learning-based LV2 DeepAFx plug-in framework [17] she regularly uses it to control her DAW and to introduce several effects. Although she always fine-tunes the work manually, the use of the framework has clearly improved her schedule. Kyra would love to have a device with an embedded version of these plug-ins for live performances.

2.4 Music generator scenarios

In this subsection, we present two scenarios that rely on the use of different AI-based music generators.

2.4.1 On hold scenario

Peter has a small online seller business with a telephone customer service line. He wants some copyright-free music to keep the costumer on hold while an agent can handle their call. He wants the music to change according to the expected waiting time, the time of the day, and other circumstances.

Peter has heard about AI-based music generation technology [5] and after searching online decides to select some compositions made using AIVA and computoser [18]. Peter consults with his guitar player friend Saul to help him decide which parameters would be best for the different music fragments that he wants for the customer service line. An automated controller dynamically changes the generator parameters to create the desired result.

Peter would like to be able to estimate the emotional state of the client [6] and change the music accordingly; however, this is not possible in a standard phone call. When clients use the customer service app, the music changes according to their comments [19]. All the generators in this scenario produce symbolic music in midi format. This format is suitable for instrumental music and produces results of a quality that can be adequate for the proposed scenario.

2.4.2 Singing elevator scenario

Mia is a Design Engineer for a large elevator company. In their latest models, the elevators are fitted with a screen that mainly provides news and weather information. Mia wants to have copyright-free background songs while the elevator is in use.

After studying several alternatives, Mia decides to generate the songs dynamically based on the characteristics of the building (residential, commercial, neighborhood, etc.). To generate the songs, she uses the OpenAi Jukebox generator [20] and updates the sons on a regular basis. The entire selection of songs according to the different situations is performed by the elevator media controller, which can also be considered a musical thing.

This scenario uses a nonsymbolic direct audio music generator. This type of generator is much less common than the symbolic alternatives, but the results are becoming acceptable by final users in the last years.

2.5 Music recommendation device scenarios

2.5.1 Emotiwatch scenario

Sam is a sports and music fan. Every morning he runs for an hour. While running, Sam likes to listen to music. His musical choices clearly depend on his mood. For years, Sam has selected his songs directly, but he would prefer, at least sometimes, that his smartwatch would do the selection for him. It is well known [21, 22] that emotional states and stress can be predicted using AI technology from physiological indicators. These are mainly electro-dermal activity (EDA), heart rate variability (HRV), and to a smaller extent, peripheral oxygen saturation (Spo2). Several wearable devices, including smartwatches such as Fitbit charge 2 or Sense [22] or research-oriented Empatica E4 wristband, are capable of measuring at least a subset of these parameters.

Sam finds an app for his watch [23] that selects music based on his mood. The watch, which was already a musical thing, becomes an AI-assisted musical device and lets Sam keep his mind on running.

2.5.2 iClock scenario

Jane, like a great part of the population in many countries, has been having lack of sleep problems for a long time. The relationships between sleep disorders and anxiety, depression, overweight, and diabetes are well known by the medical community [24]. As part of her treatment, her psychologist tells Jane that some new devices could possibly help restore her sleep quality. Among these devices, Jane finds iClock, a new device that monitors her sleep, using Jane's smartwatch, and modifies her wake up routines taking into account her schedule needs, the sleep monitoring data, and an estimation of her emotional state. Among the different aspects that iClock controls is the selection and modification of the melodies according to the selected waking up routine. Thus, iClock is, among other things, an AI-assisted musical device,

Following her therapist recommendations, including the use of iClock, Jane's sleep patterns improve, which in turn is clearly reflected in an improvement of her quality of life.

2.6 Feedback device scenarios

2.6.1 RumbleRumble scenario

Gina has a moderate hearing problem. She likes to go to concerts with friends. However, she feels that she is losing an important part of the information. Recently, she learned about the existence of the Subpac backpack [25] that uses haptics, interoception, and bone conduction to deliver bass sensation to even profoundly deaf users. Although the current version of the device requires an external computer to run the software, Gina is using an experimental version that runs in an embedded controller, thus making the Subpac a personal feedback AIME.

2.6.2 MagicShoes scenario

Peter has a problem with his weigh. He has tried several solutions, but none seem to work well for him. He has even tried game-based approaches [26] with little success. Peter is very fond of music, and he hears from a musician friend of the existence of a wearable device that uses sounds to promote sport activity and to change your own body perception. He starts using MagicShoes [27] and finally finds a way to help him reduce weight in a fun way that adequately fits his tastes and habits. A future update includes machine learning capabilities so that the device selects music based on the user preferences.

2.6.3 Let there be light scenario

Nico really likes to go to rock music performances. He especially loves when people start following music with their lighters. In some recent concerts, this has even improved due to new musical device technology. When Nico went to his last concert,

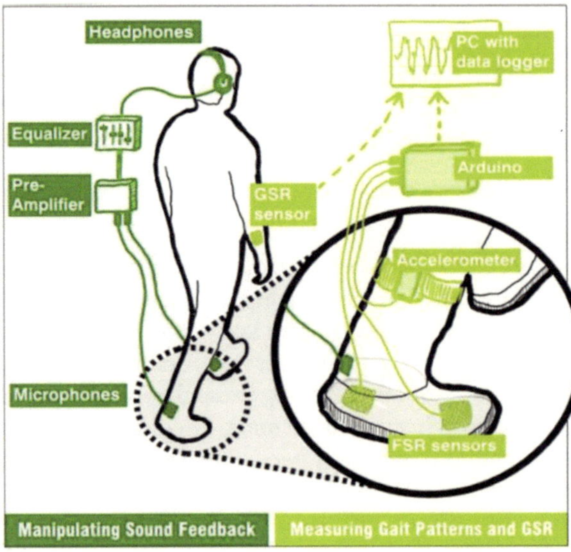

Figure 2.
MagicShoes prototype.

he was given a PixMob-led wristband. These devices have a set of preprogramed effects that are triggered usually by a human operator. Nevertheless, the possibility of an AI-based controller that decides which effect to apply according to both the concert and the carrier circumstances is currently perfectly feasible. In this way, the wristband will become an AIME (**Figure 2**).

2.7 Educational scenarios

2.7.1 Teach and play scenario: again

The Teach and Play Scenario presented in Subsection 1.2 is also clearly an educational Musical Device scenario and could have been presented in this subsection as well.

2.7.2 The magiFlute Scenario

John is 13 years old and has a moderate learning disability. His music teacher recommends that he use a new accessible digital musical instrument (ADMI) [28] known as the MagiFlute. This instrument is an Electronic Wind Instrument [EWI] [26], which is similar to a recorder, but does not produce sound directly. It has sensors for wind and touch pressure and controls a synthesizer through an embedded deep learning system. It also uses John's iPad to help him remember what to play and how to play. It even has the possibility of automatically correcting what John is playing when configured in this way. With the magiFlute, John participates in the school band and is becoming a better standard recorder player every day (**Figure 3**).

2.7.3 Magic flute scenario

In our last scenario, we use the term magiFlute for our proposed instrument, as its housemate, the "Magic Flute," is a completely different existing ADMI, which is an

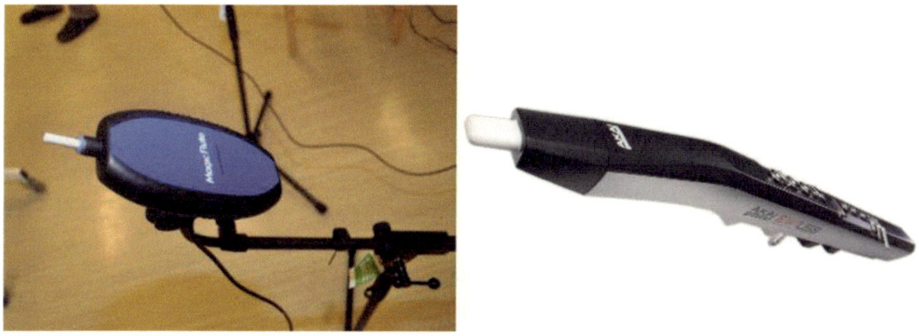

Figure 3.
Magic Flute and typical EW.

EWI that is controlled by very small head movements [27]. This instrument is played by Ellen, who has a spinal cord injury as a result of a motorcycle accident.

2.8 Processing architecture

In this work, we propose a generic architecture for the design of musical devices. This architecture is based on a multilayered approach. The proposed layers are structured as follows:

- User stimuli capture and processing layer.
- Embedded learning Layer
- Music adaptation layer
- Music production layer
- User feedback layer

The block diagram for this proposed architecture is shown in **Figure 4**.

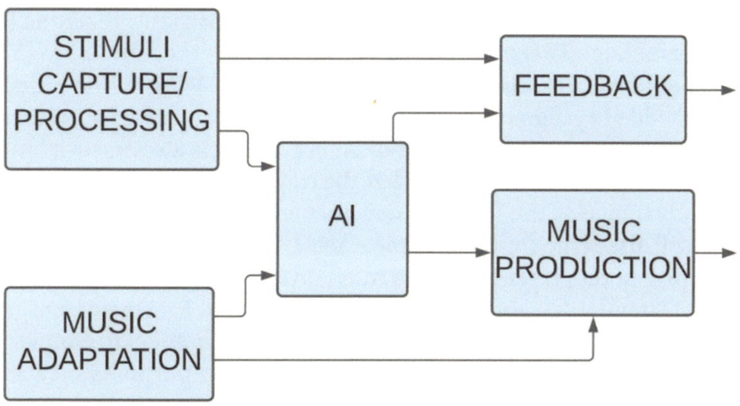

Figure 4.
Generic AIME Block diagram.

After presenting the methodology used to test the theories discussed in the Introduction, the results will be detailed in the next section.

3. Results

In this section, we present a possible implementation for some of the devices proposed in the scenarios. In this way, we will verify the suitability of the proposed generic architecture and, thus, the usefulness of the framework presented.

4. Scenario implementation

We will briefly describe possible implementations of TherAImin. This implementation is presented to show that the proposed framework provides a usable foundation for building AI-assisted devices and describing them in a systematic manner.

We think that even though we do not present a device for each of the possible categories, the difference between the selected AIMEs is wide enough to show that, in principle, any AIME can be implemented using the framework.

4.1 TherAImin

As discussed in Section 1.2, the Theremin is an instrument with two antennas that is controlled by the player without touch interaction. The block diagram of the Theremin is shown in **Figure 5**. The TherAIMin is an AI-assisted variation of the original instrument, where hand gestures are used to control the timbre.

Although we could have implemented TherAImin without antennas using, e.g., Mediapipe Handpose [29], we have decided to be more faithful to the original instrument and thus use the [30], which provides a versatile Theremin implementation with Pitch and Volume outputs.

Thus, openTheremin antennas act as part of the user-stimulus capture layer. The other part of this layer is a camera that is used to capture the user's hand gesture.

We will interface openTheramin using a Raspberry Pi board with an RPI-GP90 pulse signal IO hat. This is part of the stimulus adaptation layer. The other part of this layer is made up of the video interface already available in the raspberry pi.

The embedded learning layer is built using Google's Teachable Machine [31] accelerated with a Coral Edge TPU accelerator. The approach is very similar to [32] where a machine that can be trained is used to recognize objects. With this approach, the accelerated embedded system classifies the gestures in the number of trained classes. It is important to keep the gesture classes different, and it is also essential to train a wide class of gestures and other images that the camera may see in the background class [33]. An advantage of the TherAImin is that when the AI system makes a wrong decision, this will affect the timbre and the effects, but not the volume and the pitch.

The sound production layer is implemented on raspberry pi using sonic PI [34]. The selection of sound pitch and volume is done by a small Processing program that produces OSC [35]. Open Sound Control (OSC) is a protocol to connect sound synthesizers, computers, and other multimedia devices for purposes such as musical performance or show control. Many music-related software tools, including sonic PI, support the OSC protocol. The OSC protocol uses UDP (or TCP) packets and thus can run either in a single embedded system or be distributed over a network.

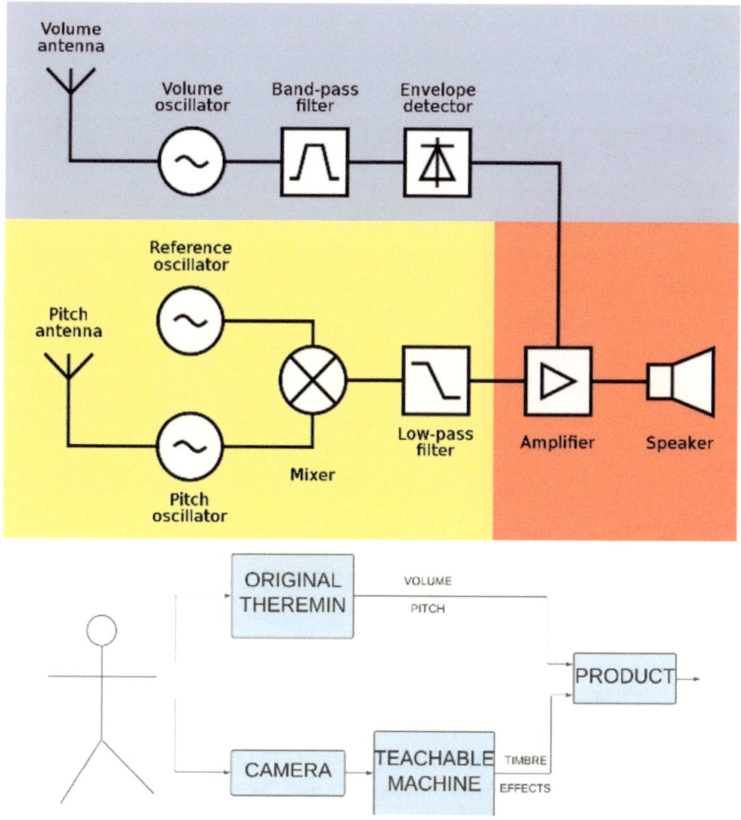

Figure 5.
Block diagram of the Theremin ad the TherAImin.

The selection of timbre and effects is done by the learning layer as a function of the gestures and sent to the generation layer using OSC. In this way, TherAImin produces audio as a function of the hand positions captured by the antennas and the gestures captured by the camera and recognized by the teachable machine.

The TherAImin, as an extension of the Theremin, can be considered in the augmented musical instrument class in the FAIME taxonomy.

5. Discussion

The approach used to implement TherAImin could be used for simpler and more complex devices. As an example, the "Singing Elevator" can use the presence, temperature, humidity and noise sensors and additional information from the Internet as user stimuli. Using aggregation of estimators this process can be further improved [36]. Using a simple learning layer (local or not), it will decide from which category it should retrieve the generated music. The production layer would be a simple player with possible adaptations to handle noise in the cabin or other issues.

As a further example, the emotiWatch uses the wearable device sensors as a stimuli layer, preprocesses them in the stimuli adaptation layer, estimates the user's emotional state, and selects the music in the learning layer and outputs the music through a player in the production layer. The same approach can be used to start the design of any AIME.

It is clear that many other approaches could have been proposed, but FAIME is simple and gives clear insights into the musical device design process.

6. Conclusions

In this work, we have presented a useful framework for the classification, understanding, and design of AI-assisted musical devices.

We have shown a very wide range of devices that can fall into this category including such different things as accessible instruments for disabled musicians or alarm clocks to help people with sleeping disorders.

We have presented a quite detailed implementation of a variation of a successful musical instrument designed in the 1920s, the Theremin. Our augmentation allows the player to select timbre and effects in real time through hand gestures but also helps to keep the look and feel of the original instrument if it is played with open hands.

We also included a short description of the design of other AIMEs to show the usefulness of the framework.

In future work, we will evaluate the user experience of TherAImin with musicians and study possible modifications for performers with disabilities using the powerful embedded intelligent system.

Acknowledgements

This work was supported by the NICO project (PID2019-105455GB-C31) from Ministerio de Ciencia, Innovación y Universidades (Spanish Government) and by DAFNE (US-1381619) Consejería de Economía y Conocimiento (Junta de Andalucia).

Author details

Miguel Civit[1*], Luis Muñoz-Saavedra[2], Francisco Cuadrado[1], Charles Tijus[3] and María José Escalona[4]

1 Loyola University, Seville, Spain

2 E.T.S. Ingeniería Informática, University of Seville, Seville, Spain

3 University of Paris, Paris, France

4 Computer Languages and Systems Department, ETSII, University of Seville, Seville, Spain

*Address all correspondence to: mcivit@uloyola.es

References

[1] Turchet L, Fischione C, Essl G, Keller D, Barthet M. Internet of musical things: Vision and challenges. IEEE Access. 2018;**6**:61994-62017

[2] Clark D, Westin F, Girouard A. iSNoW: User perceptions of an interactive social novelty wearable. In: Adjunct Proceedings of the 2019 ACM International Joint Conference on Pervasive and Ubiquitous Computing and Proceedings of the 2019 ACM International Symposium on Wearable Computers. 2019. pp. 268-271

[3] Fiebrink R, Sonami L. Reflections on Eight Years of Instrument Creation with Machine Learning. Goldsmiths, University of London; 2020

[4] Buehler-McWilliams K, Murray RE. The monochord in the medieval and modern classrooms. Journal of Music History Pedagogy. 2013;**3**:151-172

[5] Briot JP, Hadjeres G, Pachet FD. Deep Learning Techniques for Music Generation. Springer; 2020

[6] Muñoz-Saavedra L, Luna-Perejón F, Civit-Masot J, Miró-Amarante L, Civit A, Domínguez-Morales M. Affective state assistant for helping users with cognition disabilities using neural networks. Electronics. 2020;**9**:1843

[7] Rahaman T. Smart things are getting smarter: An introduction to the internet of behavior. Medical Reference Services Quarterly. 2022;**41**:110-116

[8] Jordà S. Instruments and players: Some thoughts on digital lutherie. Journal of New Music Research. 2004;**33**:321-341

[9] Harrison J. Instruments and Access: The Role of Instruments in Music and Disability [Ph.D. dissertation]. Queen Mary University of London; 2020

[10] Dieckmann M. EMG/Motion Capture-Based Accessible Music Interfaces for Rehabilitation. 2020

[11] Theremin LS, Petrishev O. The design of a musical instrument based on cathode relays. Leonardo Music Journal. 1996;**6**:49-50

[12] McAdams S, Giordano BL. The perception of musical timbre. In: The Oxford Handbook of Music Psychology. Oxford Academic; 2009. pp. 72-80

[13] Wright A, Damskägg EP, Juvela L, Välimäki V. Real-time guitar amplifier emulation with deep learning. Applied Sciences. 2020;**10**:766

[14] Civit-Masot J, Luna-Perejón F, Corral JMR, Domínguez-Morales M, Morgado-Estévez A, Civit A. A study on the use of Edge TPUs for eye fundus image segmentation. Engineering Applications of Artificial Intelligence. 2021;**104**:104384

[15] Mårtensson B. The Timbral and Quality Affect from Pitch Correction Software on a Recorded Vocal Performance [Dissertation]. 2022. Retrieved from: http://urn.kb.se/resolve?urn=urn:nbn:se:ltu:diva-90744

[16] Wager S, Tzanetakis G, CI W, Kim M. Deep autotuner: A pitch correcting network for singing performances. In: ICASSP 2020-2020 IEEE International Conference on Acoustics, Speech and Signal Processing (ICASSP). 2020. pp. 246-250

[17] Martinez Ramirez MA, Wang O, Smaragdis P, Bryan NJ. Differentiable signal processing with black-box audio effects. In: IEEE International Conference on Acoustics, Speech and Signal Processing (ICASSP). IEEE; 2021

[18] Bozhanov B. Computoser-Rule-Based, Probability-Driven Algorithmic Music Composition. arXiv preprint arXiv:1412.3079. 2014

[19] Salas J. Generating music from literature using topic extraction and sentiment analysis. IEEE Potentials. 2018;**37**:15-18

[20] Dhariwal P, Jun H, Payne C, Kim JW, Radford A, Sutskever I. Jukebox: A Generative Model for Music. arXiv preprint arXiv:2005.00341. 2020

[21] Assabumrungrat R et al. Ubiquitous affective computing: A review. IEEE Sensors Journal. 1 Feb 2022;**22**(3):1867-1881. DOI: 10.1109/JSEN.2021.3138269

[22] Williams SH. A Validation Study: Fitbit Charge 2 Heart Rate Measurement at Rest and During Cognitive-Emotional Stressors. 2021

[23] Linger O. Designing a User-Centered Music Experience for the Smartwatch [Dissertation]. 2018. Retrieved from: http://urn.kb.se/resolve?urn=urn:nbn:se:kth:diva-231061

[24] Staner L. Sleep and anxiety disorders. Dialogues in Clinical Neuroscience. 2003;**5**(3):249-258. DOI: 10.31887/DCNS.2003.5.3/lstaner

[25] Schmitz A, Holloway C, Cho Y. Hearing through vibrations: Perception of musical emotions by profoundly deaf people. arXiv preprint arXiv:2012.13265. 2020

[26] Snyder J. The birl: Adventures in the development of an electronic wind instrument. In: Musical Instruments in the 21st Century. Springer; 2017. pp. 181-205

[27] Davanzo N, Avanzini F. Experimental evaluation of three interaction channels for accessible digital musical instruments. In: International Conference on Computers Helping People with Special Needs. 2020. pp. 437-445

[28] Frid E. Accessible digital musical instruments—a review of musical interfaces in inclusive music practice. Multimodal Technologies and Interaction (MDPI). 2019;**3**

[29] Sung G, Sokal K, Uboweja E, Bazarevsky V, Baccash J, Bazavan EG, et al. On-device Real-time Hand Gesture Recognition. arXiv preprint arXiv:2111.00038. 2021

[30] GaudiLabs. Open Theremin - Open Source Hardware Project. 2022. Available from: https://github.com/GaudiLabs/OpenTheremin_V3

[31] Carney M, Webster B, Alvarado I, Phillips K, Howell N, Griffith J, et al. Teachable machine: Approachable Web-based tool for exploring machine learning classification. In: Extended Abstracts of the 2020 CHI Conference on Human Factors in Computing Systems. 2020. pp. 1-8

[32] Tyka M. Embedded Teachable Machine. April 2019. Available from: https://teachablemachine.withgoogle.com/

[33] Muñoz-Saavedra L, Civit-Masot J, Luna-Perejón F, Domínguez-Morales M, Civit A. Does two-class training extract real features? a COVID-19 case study. Applied Sciences. 2021;**11**:1424

[34] Aaron S, Blackwell AF, Burnard P. The development of Sonic Pi and its use in educational partnerships: Co-creating pedagogies for learning computer programming. Journal of Music, Technology & Education. 2016;**9**:75-94

[35] Wright M. OpenSound Control Specification. UC Berkeley: Center for New Music and Audio Technologies; 2002

[36] El Ghali K, El Ghali A, Tijus C. Multimodal Automatic Tagging of Music Titles using Aggregation of Estimators. MediaEval; 2012